PERT MATH STUDY GUIDE 2020-2021

A Comprehensive Review and Step-By-Step Guide to Preparing for the PERT Math

By

Reza Nazari & Ava Ross

About Effortless Math Education Inc.

Effortless Math Education operates the www.effortlessmath.com website, which prepares and publishes Test prep and Mathematics learning resources. Effortless Math authors' team strives to prepare and publish the best quality Mathematics learning resources to make learning Math easier for all. We Help Students Learn to Love Mathematics.

All inquiries should be addressed to:

info@effortlessMath.com

www.EffortlessMath.com

ISBN: 978-1-64612-410-7

Published by: Effortless Math Education Inc.

www.EffortlessMath.com

Visit www.EffortlessMath.com

for Online Math Practice

Welcome to

PERT Math Prep
2021

Thank you for choosing Effortless Math for your PERT Math test preparation and congratulations on making the decision to take the PERT test! It's a remarkable move you are taking, one that shouldn't be diminished in any capacity. That's why you need to use every tool possible to ensure you succeed on the test with the highest possible score, and this extensive study guide is one such tool.

If math has never been a strong subject for you, **don't worry**! This book will help you prepare for (and even ACE) the PERT test's math section. As test day draws nearer, effective preparation becomes increasingly more important. Thankfully, you have this comprehensive study guide to help you get ready for the test. With this guide, you can feel confident that you will be more than ready for the PERT Math test when the time comes.

First and foremost, it is important to note that this book is a study guide and not a textbook. It is best read from cover to cover. Every lesson of this "self-guided math book" was carefully developed to ensure that you are making the most effective use of your time while preparing for the test. This up-to-date guide reflects the 2021 test guidelines and will put you on the right track to hone your math skills, overcome exam anxiety, and boost your confidence, so that you can have your best to succeed on the PERT Math test.

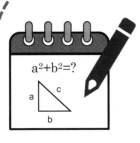

This study guide will:

☑ Explain the format of the PERT Math test.

☑ Describe specific test-taking strategies that you can use on the test.

☑ Provide PERT Math test-taking tips.

☑ Review all PERT Math concepts and topics you will be tested on.

☑ Help you identify the areas in which you need to concentrate your study time.

☑ Offer exercises that help you develop the basic math skills you will learn in each section.

☑ Give **2 realistic and full-length practice tests** (featuring new question types) with detailed answers to help you measure your exam readiness and build confidence.

This resource contains everything you will ever need to succeed on the PERT Math test. You'll get in-depth instructions on every math topic as well as tips and techniques on how to answer each question type. You'll also get plenty of practice questions to boost your test-taking confidence.

In addition, in the following pages you'll find:

➢ **How to Use This Book Effectively** – This section provides you with step-by-step instructions on how to get the most out of this comprehensive study guide.

➢ **How to study for the PERT Math Test** – A six-step study program has been developed to help you make the best use of this book and prepare for your PERT Math test. Here you'll find tips and strategies to guide your study program and help you understand PERT Math and how to ace the test.

➢ **PERT Math Review** – Learn everything you need to know about the PERT Math test.

➢ **PERT Math Test-Taking Strategies** – Learn how to effectively put these recommended test-taking techniques into use for improving your PERT Math score.

➢ **Test Day Tips** – Review these tips to make sure you will do your best when the big day comes.

➢

Effortless Math's PERT Online Center

Effortless Math Online PERT Center offers a complete study program, including the following:

✓ Step-by-step instructions on how to prepare for the PERT Math test

✓ Numerous PERT Math worksheets to help you measure your math skills

✓ Complete list of PERT Math formulas

✓ Video lessons for all PERT Math topics

✓ Full-length PERT Math practice tests

✓ And much more...

No Registration Required.

Visit **EffortlessMath.com/PERT** to find your online PERT Math resources.

How to Use This Book Effectively

Look no further when you need a study guide to improve your math skills to succeed on the math portion of the PERT test. Each chapter of this comprehensive guide to the PERT Math will provide you with the knowledge, tools, and understanding needed for every topic covered on the test.

It's imperative that you understand each topic before moving onto another one, as that's the way to guarantee your success. Each chapter provides you with examples and a step-by-step guide of every concept to better understand the content that will be on the test. To get the best possible results from this book:

➢ **Begin studying long before your test date**. This provides you ample time to learn the different math concepts. The earlier you begin studying for the test, the sharper your skills will be. Do not procrastinate! Provide yourself with plenty of time to learn the concepts and feel comfortable that you understand them when your test date arrives.

➢ **Practice consistently**. Study PERT Math concepts at least 20 to 30 minutes a day. Remember, slow and steady wins the race, which can be applied to preparing for the PERT Math test. Instead of cramming to tackle everything at once, be patient and learn the math topics in short bursts.

➢ Whenever you get a math problem wrong, **mark it off, and review it later** to make sure you understand the concept.

➢ Start each session by **looking over the previous material.**

➢ Once you've reviewed the book's lessons, **take a practice test at the back of the book** to gauge your level of readiness. Then, review your results. Read detailed answers and solutions for each question you missed.

➢ **Take another practice test** to get an idea of how ready you are to take the actual exam. Taking the practice tests will give you the confidence you need on test day. Simulate the PERT testing environment by sitting in a quiet room free from distraction. Make sure to clock yourself with a timer.

How to Study for the PERT Math Test

Studying for the PERT Math test can be a really daunting and boring task. What's the best way to go about it? Is there a certain study method that works better than others? Well, studying for the PERT Math can be done effectively. The following six-step program has been designed to make preparing for the PERT Math test more efficient and less overwhelming.

Step **1** - Create a study plan
Step **2** - Choose your study resources
Step **3** - Review, Learn, Practice
Step **4** - Learn and practice test-taking strategies
Step **5** - Learn the PERT Test format and take practice tests
Step **6** - Analyze your performance

STEP 1: Create a Study Plan

It's always easier to get things done when you have a plan. Creating a study plan for the PERT Math test can help you to stay on track with your studies. It's important to sit down and prepare a study plan with what works with your life, work, and any other obligations you may have. Devote enough time each day to studying. It's also a great

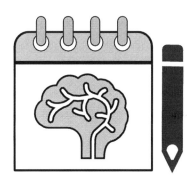

idea to break down each section of the exam into blocks and study one concept at a time.

It's important to understand that there is no "right" way to create a study plan. Your study plan will be personalized based on your specific needs and learning style. Follow these guidelines to create an effective study plan for your PERT Math test:

★ **Analyze your learning style and study habits** – Everyone has a different learning style. It is essential to embrace your individuality and the unique way you learn. Think about what works and what doesn't work for you. Do you prefer PERT Math prep books or a combination of textbooks and video lessons? Does it work better for you if you study every night for thirty minutes or is it more effective to study in the morning before going to work?

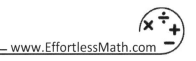

★ **Evaluate your schedule** – Review your current schedule and find out how much time you can consistently devote to PERT Math study.

★ **Develop a schedule** – Now it's time to add your study schedule to your calendar like any other obligation. Schedule time for study, practice, and review. Plan out which topic you will study on which day to ensure that you're devoting enough time to each concept. Develop a study plan that is mindful, realistic, and flexible.

★ **Stick to your schedule** – A study plan is only effective when it is followed consistently. You should try to develop a study plan that you can follow for the length of your study program.

★ **Evaluate your study plan and adjust as needed** – Sometimes you need to adjust your plan when you have new commitments. Check in with yourself regularly to make sure that you're not falling behind in your study plan. Remember, the most important thing is sticking to your plan. Your study plan is all about helping you be more productive. If you find that your study plan is not as effective as you want, don't get discouraged. It's okay to make changes as you figure out what works best for you.

STEP 2: Choose Your Study Resources

There are numerous textbooks and online resources available for the PERT Math test, and it may not be clear where to begin. Don't worry! This study guide provides everything you need to fully prepare for your PERT Math test. In addition to the book content, you can also use Effortless Math's online resources. (video lessons, worksheets, formulas, etc.)

Simply visit EffortlessMath.com/PERT to find your online PERT Math resources.

STEP 3: Review, Learn, Practice

This PERT Math study guide breaks down each subject into specific skills or content areas. For instance, the percent concept is separated into different topics–percent calculation, percent increase and decrease, percent problems, etc. Use this book to help you go over all key math concepts and topics on the PERT Math test.

As you read each chapter, take notes or highlight the concepts you would like to go over again in the future. If you're unfamiliar with a topic or something is difficult for you, do additional research on it. For each math topic, plenty of instructions, step-by-step guides, and examples are provided to ensure you get a good grasp of the material. You can also find video lessons on the Effortless Math website for each PERT Math concept. Quickly review the topics you do understand to get a brush-up of the material. Be sure to do the practice questions provided at the end of every chapter to measure your understanding of the concepts.

STEP 4: Learn and Practice Test-taking Strategies

In the following sections, you will find important test-taking strategies and tips that can help you earn extra points. You'll learn how to think strategically and when to guess if you don't know the answer to a question. Using PERT Math test-taking strategies and tips can help you raise your score and do well on the test. Apply test taking strategies on the practice tests to help you boost your confidence.

STEP 5: Learn the PERT Test Format and Take Practice Tests

The PERT *Test Review* section provides information about the structure of the PERT test. Read this section to learn more about the PERT test structure, different test sections, the number of questions in each section, and the section time limits. When you have a prior understanding of the test format and different types of PERT Math questions, you'll feel more confident when you take the actual exam.

Once you have read through the instructions and lessons and feel like you are ready to go – take advantage of both of the full-length PERT Math practice tests available in this study guide. Use the practice tests to sharpen your skills and build confidence. The PERT Math practice tests offered at the end of the book are formatted similarly to the actual PERT Math test. When you take each practice test, try to simulate actual testing conditions. To take the practice tests, sit in a quiet space, time yourself, and work through as many of the questions as time allows. The practice tests are followed by detailed answer explanations to help you find your weak areas, learn from your mistakes, and raise your PERT Math score.

STEP 6: Analyze Your Performance

After taking the practice tests, look over the answer keys and explanations to learn which questions you answered correctly and which you did not. Never be discouraged if you make a few mistakes. See them as a learning opportunity. This will highlight your strengths and weaknesses.

You can use the results to determine if you need additional practice or if you are ready to take the actual PERT Math test.

Looking for more?

Visit EffortlessMath.com/PERT to find hundreds of PERT Math worksheets, video tutorials, practice tests, PERT Math formulas, and much more.

Or scan this QR code.

No Registration Required.

PERT Test Review

The Postsecondary Education Readiness Test, is known as the PERT, is a test to determine the appropriate level of college course work for an incoming student. In essence, it is a broad and quick assessment of students' academic abilities.

The PERT test consists of three multiple-choice separate exams:

- Mathematics
- Reading
- Writing

The PERT test is a Computer Adaptive Test (CAT). It means that if the correct answer is chosen, the next question will be harder. If the answer given is incorrect, the next question will be easier. This also means that once an answer is selected on the CAT it cannot be changed.

The mathematics portion of the PERT test contains 30 multiple-choice questions. The test covers data analysis, geometry, and algebra on both intermediate and basic levels. The topics on the math test includes:

- linear equations, linear inequalities, literal equation, and quadratic formulas
- simultaneous linear equations with two variables
- evaluating algebraic expressions
- translating between lines and inspecting equations on coordinate planes
- dividing by binomials and monomials
- adding, subtracting, multiplying, dividing, simplifying and factoring polynomial

Students are not allowed to use calculator when taking a PERT assessment. A pop-up calculator is embedded in the test for some questions. Scores on the PERT test range from 50 to 150. On the PERT mathematics exam, you will be placed in lower-level developmental education if you score between 50-95. If your score is between 96 and 113, then you will be placed in higher level developmental education. A score of 114 to 122 will enable you to be place in Intermediate Algebra (MAT 1033). Score of 123 or higher will allow you to take College Algebra or higher (MAC 1105).

PERT Math Test-Taking Strategies

Here are some test-taking strategies that you can use to maximize your performance and results on the PERT Math test.

#1: USE THIS APPROACH TO ANSWER EVERY PERT MATH QUESTION

- Review the question to identify keywords and important information.
- Translate the keywords into math operations so you can solve the problem.
- Review the answer choices. What are the differences between answer choices?
- Draw or label a diagram if needed.
- Try to find patterns.
- Find the right method to answer the question. Use straightforward math, plug in numbers, or test the answer choices (backsolving).
- Double-check your work.

#2: USE EDUCATED GUESSING

This approach is applicable to the problems you understand to some degree but cannot solve using straightforward math. In such cases, try to filter out as many answer choices as possible before picking an answer. In cases where you don't have a clue about what a certain problem entails, don't waste any time trying to eliminate answer choices. Just choose one randomly before moving onto the next question.

As you can ascertain, direct solutions are the most optimal approach. Carefully read through the question, determine what the solution is using the math you have learned before, then coordinate the answer with one of the choices available to you. Are you stumped? Make your best guess, then move on.

Don't leave any fields empty! Even if you're unable to work out a problem, strive to answer it. Take a guess if you have to. You will not lose points by getting an answer wrong, though you may gain a point by getting it correct!

#3 : BALLPARK

A ballpark answer is a rough approximation. When we become overwhelmed by calculations and figures, we end up making silly mistakes. A decimal that is moved by one unit can change an answer from right to wrong, regardless of the number of steps that you went through to get it. That's where ballparking can play a big part.

If you think you know what the correct answer may be (even if it's just a ballpark answer), you'll usually have the ability to eliminate a couple of choices. While answer choices are usually based on the average student error and/or values that are closely tied, you will still be able to weed out choices that are way far afield. Try to find answers that aren't in the proverbial ballpark when you're looking for a wrong answer on a multiple-choice question. This is an optimal approach to eliminating answers to a problem.

#4 : BACKSOLVING

All questions on the PERT Math test will be in multiple-choice format. Many test-takers prefer multiple-choice questions, as at least the answer is right there. You'll typically have four answers to pick from. You simply need to figure out which one is correct. Usually, the best way to go about doing so is "backsolving."

As mentioned earlier, direct solutions are the most optimal approach to answering a question. Carefully read through a problem, calculate a solution, then correspond the answer with one of the choices displayed in front of you. If you can't calculate a solution, your next best approach involves "backsolving."

When backsolving a problem, contrast one of your answer options against the problem you are asked, then see which of them is most relevant. More often than not, answer choices are listed in ascending or descending order. In such cases, try out the choices B or C. If it's not correct, you can go either down or up from there.

#5 : PLUGGING IN NUMBERS

"Plugging in numbers" is a strategy that can be applied to a wide range of different math problems on the PERT Math test. This approach is typically used to simplify a challenging question so that it is more understandable. By using the strategy carefully, you can find the answer without too much trouble.

The concept is fairly straightforward—replace unknown variables in a problem with certain values. When selecting a number, consider the following:

- Choose a number that's basic (just not too basic). Generally, you should avoid choosing 1 (or even 0). A decent choice is 2.

- Try not to choose a number that is displayed in the problem.

- Make sure you keep your numbers different if you need to choose at least two of them.

- More often than not, choosing numbers merely lets you filter out some of your answer choices. As such, don't just go with the first choice that gives you the right answer.

- If several answers seem correct, then you'll need to choose another value and try again. This time, though, you'll just need to check choices that haven't been eliminated yet.

- If your question contains fractions, then a potential right answer may involve either an LCD (least common denominator) or an LCD multiple.

- 100 is the number you should choose when you are dealing with problems involving percentages.

PERT Mathematics Test – Daytime Tips

After practicing and reviewing all the math concepts you've been taught, and taking some PERT mathematics practice tests, you'll be prepared for test day. Consider the following tips to be extra-ready come test time.

Before Your Test

What to do the night before:

- **Relax!** One day before your test, study lightly or skip studying altogether. You shouldn't attempt to learn something new, either. There are plenty of reasons why studying the evening before a big test can work against you. Put it this way–a marathoner wouldn't go out for a sprint before the day of a big race. Mental marathoners–such as yourself–should not study for any more than one hour 24 hours before a PERT test. That's because your brain requires some rest to be at its best. The night before your exam, spend some time with family or friends, or read a book.

- **Avoid bright screens** - You'll have to get some good shuteye the night before your test. Bright screens (such as the ones coming from your laptop, TV, or mobile device) should be avoided altogether. Staring at such a screen will keep your brain up, making it hard to drift asleep at a reasonable hour.

- **Make sure your dinner is healthy** - The meal that you have for dinner should be nutritious. Be sure to drink plenty of water as well. Load up on your complex carbohydrates, much like a marathon runner would do. Pasta, rice, and potatoes are ideal options here, as are vegetables and protein sources.

- **Get your bag ready for test day** - The night prior to your test, pack your bag with your stationery, admissions pass, ID, and any other gear that you need. Keep the bag right by your front door.

- **Make plans to reach the testing site** - Before going to sleep, ensure that you understand precisely how you will arrive at the site of the test. If parking is something you'll have to find first, plan for it. If you're dependent on public transit, then review the schedule. You should also make sure that the train/bus/subway/streetcar you use will be running. Find out about road closures as well. If a parent or friend is accompanying you, ensure that they understand what steps they have to take as well.

The Day of the Test

- **Get up reasonably early, but not too early.**

- **Have breakfast** - Breakfast improves your concentration, memory, and mood. As such, make sure the breakfast that you eat in the morning is healthy. The last thing you want to be is distracted by a grumbling tummy. If it's not your own stomach making those noises, another test taker close to you might be instead. Prevent discomfort or embarrassment by consuming a healthy breakfast. Bring a snack with you if you think you'll need it.

- **Follow your daily routine** - Do you watch Good Morning America each morning while getting ready for the day? Don't break your usual habits on the day of the test. Likewise, if coffee isn't something you drink in the morning, then don't take up the habit hours before your test. Routine consistency lets you concentrate on the main objective–doing the best you can on your test.

- **Wear layers** - Dress yourself up in comfortable layers. You should be ready for any kind of internal temperature. If it gets too warm during the test, take a layer off.

- **Get there on time** - The last thing you want to do is get to the test site late. Rather, you should be there 45 minutes prior to the start of the test. Upon your arrival, try not to hang out with anybody who is nervous. Any anxious energy they exhibit shouldn't influence you.

- **Leave the books at home** - No books should be brought to the test site. If you start developing anxiety before the test, books could encourage you to do some last-minute studying, which will only hinder you. Keep the books far away–better yet, leave them at home.

- **Make your voice heard** - If something is off, speak to a proctor. If medical attention is needed or if you'll require anything, consult the proctor prior to the start of the test. Any doubts you have should be clarified. You should be entering the test site with a state of mind that is completely clear.

- **Have faith in yourself** - When you feel confident, you will be able to perform at your best. When you are waiting for the test to begin, envision yourself receiving an outstanding result. Try to see yourself as someone who knows all the answers, no matter what the questions are. A lot of athletes tend to use this technique—particularly before a big competition. Your expectations will be reflected by your performance.

During your test

- **Be calm and breathe deeply** - You need to relax before the test, and some deep breathing will go a long way to help you do that. Be confident and calm. You got this. Everybody feels a little stressed out just before an evaluation of any kind is set to begin. Learn some effective breathing exercises. Spend a minute meditating before the test starts. Filter out any negative thoughts you have. Exhibit confidence when having such thoughts.

- **Concentrate on the test** - Refrain from comparing yourself to anyone else. You shouldn't be distracted by the people near you or random noise. Concentrate exclusively on the test. If you find yourself irritated by surrounding noises, earplugs can be used to block sounds off close to you. Don't forget—the test is going to last several hours if you're taking more than one subject of the test. Some of that time will be dedicated to brief sections. Concentrate on the specific section you are working on during a particular moment. Do not let your mind wander off to upcoming or previous sections.

- **Skip challenging questions** - Optimize your time when taking the test. Lingering on a single question for too long will work against you. If you don't know what the answer is to a certain question, use your best guess and move on. That time would be better served handling the questions you can actually answer well.

- **Try to answer each question individually** - Focus only on the question you are working on. Use one of the test-taking strategies to solve the problem. If you aren't able to come up with an answer, don't get frustrated. Simply skip that question, then move onto the next one.

- **Don't forget to breathe!** Whenever you notice your mind wandering, your stress levels boosting, or frustration brewing, take a thirty-second break. Shut your eyes, drop your pencil, breathe deeply, and let your shoulders relax. You will end up being more productive when you allow yourself to relax for a moment.

■ **Optimize your breaks** - When break time comes, use the restroom, have a snack, and reactivate your energy for the subsequent section. Doing some stretches can help stimulate your blood flow.

After your test

■ **Take it easy** - You will need to set some time aside to relax and decompress once the test has concluded. There is no need to stress yourself out about what you could've said, or what you may have done wrong. At this point, there's nothing you can do about it. Your energy and time would be better spent on something that will bring you happiness for the remainder of your day.

■ **Redoing the test** - Did you pass the test? Congratulations! Your hard work paid off!

If you have failed your test, though, don't worry! The test can be retaken. In such cases, you will need to follow the retake policy. You also need to re-register to take the exam again.

Contents

Name:	Date: ....

Topic	**Simplifying Fractions**
Notes	✓ Evenly divide both the top and bottom of the fraction by $2, 3, 5, 7, \dots$ etc. ✓ Continue until you can't go any further.
Example	***Simplify*** $\frac{36}{48}$ To simplify $\frac{36}{48}$, find a number that both 36 and 48 are divisible by. Both are divisible by 12. Then: $\frac{36}{48} = \frac{36 \div 12}{48 \div 12} = \frac{3}{4}$

Your Turn!	1) $\frac{2}{18} =$	2) $\frac{22}{66} =$
	3) $\frac{12}{48} =$	4) $\frac{11}{99} =$
	5) $\frac{15}{75} =$	6) $\frac{25}{100} =$
	7) $\frac{16}{72} =$	8) $\frac{32}{96} =$
	9) $\frac{14}{77} =$	10) $\frac{60}{84} =$

Name: ....

Date: ....

Topic	Simplifying Fractions - Answers
Notes	✓ Evenly divide both the top and bottom of the fraction by $2, 3, 5, 7, \ldots$ etc. ✓ Continue until you can't go any further.
Example	*Simplify* $\frac{36}{48}$ To simplify $\frac{36}{48}$, find a number that both 36 and 48 are divisible by. Both are divisible by 12. Then: $\frac{36}{48} = \frac{36 \div 12}{48 \div 12} = \frac{3}{4}$

Your Turn!		
	1) $\frac{2}{18} = \frac{1}{9}$	2) $\frac{22}{66} = \frac{1}{3}$
	3) $\frac{12}{48} = \frac{1}{4}$	4) $\frac{11}{99} = \frac{1}{9}$
	5) $\frac{15}{75} = \frac{1}{5}$	6) $\frac{25}{100} = \frac{1}{4}$
	7) $\frac{16}{72} = \frac{2}{9}$	8) $\frac{32}{96} = \frac{1}{3}$
	9) $\frac{14}{77} = \frac{2}{11}$	10) $\frac{60}{84} = \frac{5}{7}$

| Name: .... | Date: .... |

Topic	**Adding and Subtracting Fractions**
Notes	✓ For "like" fractions (fractions with the same denominator), add or subtract the numerators and write the answer over the common denominator. ✓ Find equivalent fractions with the same denominator before you can add or subtract fractions with different denominators. ✓ Adding and Subtracting with the same denominator: $$\frac{a}{b} + \frac{c}{b} = \frac{a+c}{b} , \frac{a}{b} - \frac{c}{b} = \frac{a-c}{b}$$ ✓ Adding and Subtracting fractions with different denominators: $$\frac{a}{b} + \frac{c}{d} = \frac{ad + bc}{bd} , \frac{a}{b} - \frac{c}{d} = \frac{ad - bc}{bd}$$
Example	*Find the sum.* $\frac{3}{5} + \frac{2}{3} = \frac{(3)3+(5)(2)}{5 \times 3} = \frac{19}{15}$ *Subtract.* $\frac{4}{7} - \frac{3}{7} = \frac{1}{7}$

Your Turn!	1) $\frac{3}{5} + \frac{2}{7} =$	2) $\frac{7}{9} - \frac{4}{7} =$
	3) $\frac{4}{9} + \frac{5}{8} =$	4) $\frac{5}{8} - \frac{2}{5} =$
	5) $\frac{2}{5} + \frac{1}{6} =$	6) $\frac{2}{3} - \frac{1}{4} =$
	7) $\frac{8}{9} + \frac{5}{7} =$	8) $\frac{6}{7} - \frac{5}{9} =$

Name: ....   **Date:** ....

Topic	Adding and Subtracting Fractions - Answers
Notes	✓ For "like" fractions (fractions with the same denominator), add or subtract the numerators and write the answer over the common denominator. ✓ Find equivalent fractions with the same denominator before you can add or subtract fractions with different denominators. ✓ Adding and Subtracting with the same denominator: $$\frac{a}{b}+\frac{c}{b}=\frac{a+c}{b},\ \frac{a}{b}-\frac{c}{b}=\frac{a-c}{b}$$ ✓ Adding and Subtracting fractions with different denominators: $$\frac{a}{b}+\frac{c}{d}=\frac{ad+bc}{bd},\frac{a}{b}-\frac{c}{d}=\frac{ad-bc}{bd}$$
Example	***Find the sum.*** $\frac{3}{5}+\frac{2}{3}=\frac{(3)3+(5)(2)}{5\times3}=\frac{19}{15}$ ***Subtract.*** $\frac{4}{7}-\frac{3}{7}=\frac{1}{7}$
Your Turn!	1) $\frac{3}{5}+\frac{2}{7}=\frac{31}{35}$ 2) $\frac{7}{9}-\frac{4}{7}=\frac{13}{63}$ 3) $\frac{4}{9}+\frac{5}{8}=\frac{77}{72}$ 4) $\frac{5}{8}-\frac{2}{5}=\frac{9}{40}$ 5) $\frac{2}{5}+\frac{1}{6}=\frac{17}{30}$ 6) $\frac{2}{3}-\frac{1}{4}=\frac{5}{12}$ 7) $\frac{8}{9}+\frac{5}{7}=\frac{101}{63}$ 8) $\frac{6}{7}-\frac{5}{9}=\frac{19}{63}$

Name: …………………………………….	Date: ………………………………………

Topic	**Multiplying and Dividing Fractions**
Notes	✓ Multiplying fractions: multiply the top numbers and multiply the bottom numbers. ✓ Dividing fractions: Keep, Change, Flip Keep first fraction, change division sign to multiplication, and flip the numerator and denominator of the second fraction. Then, solve!
Examples	***Multiply.*** $\frac{2}{5} \times \frac{3}{4} =$ Multiply the top numbers and multiply the bottom numbers. $\frac{2}{5} \times \frac{3}{4} = \frac{2\times3}{5\times4} = \frac{6}{20}$, simplify: $\frac{6}{2} = \frac{6\div2}{20\div2} = \frac{3}{10}$ ***Divide.*** $\frac{2}{5} \div \frac{3}{4} =$ Keep first fraction, change division sign to multiplication, and flip the numerator and denominator of the second fraction. Then: $\frac{2}{5} \div \frac{3}{4} = \frac{2}{5} \times \frac{4}{3} = \frac{2\times4}{5\times3} = \frac{8}{15}$
Your Turn!	1) $\frac{5}{9} \times \frac{4}{7} =$ 2) $\frac{3}{5} \div \frac{2}{3} =$ 3) $\frac{2}{7} \times \frac{3}{5} =$ 4) $\frac{2}{5} \div \frac{7}{12} =$ 5) $\frac{1}{7} \times \frac{4}{9} =$ 6) $\frac{2}{9} \div \frac{3}{7} =$ 7) $\frac{2}{5} \times \frac{6}{7} =$ 8) $\frac{1}{4} \div \frac{2}{5} =$

Name:	Date: ....

Topic	Multiplying and Dividing Fractions - Answers	
Notes	✓ Multiplying fractions: multiply the top numbers and multiply the bottom numbers. ✓ Dividing fractions: Keep, Change, Flip Keep first fraction, change division sign to multiplication, and flip the numerator and denominator of the second fraction. Then, solve!	

Examples

Multiply. $\dfrac{2}{5} \times \dfrac{3}{4} =$

Multiply the top numbers and multiply the bottom numbers.

$\dfrac{2}{5} \times \dfrac{3}{4} = \dfrac{2\times3}{5\times4} = \dfrac{6}{20}$, simplify: $\dfrac{6}{2} = \dfrac{6\div2}{20\div2} = \dfrac{3}{10}$

Divide. $\dfrac{2}{5} \div \dfrac{3}{4} =$

Keep first fraction, change division sign to multiplication, and flip the numerator and denominator of the second fraction.

Then: $\dfrac{2}{5} \div \dfrac{3}{4} = \dfrac{2}{5} \times \dfrac{4}{3} = \dfrac{2\times4}{5\times3} = \dfrac{8}{15}$

Your Turn!

1) $\dfrac{5}{9} \times \dfrac{4}{7} = \dfrac{20}{63}$

2) $\dfrac{3}{5} \div \dfrac{2}{3} = \dfrac{9}{10}$

3) $\dfrac{2}{7} \times \dfrac{3}{5} = \dfrac{6}{35}$

4) $\dfrac{2}{5} \div \dfrac{7}{12} = \dfrac{24}{35}$

5) $\dfrac{1}{7} \times \dfrac{4}{9} = \dfrac{4}{63}$

6) $\dfrac{2}{9} \div \dfrac{3}{7} = \dfrac{14}{27}$

7) $\dfrac{2}{5} \times \dfrac{6}{7} = \dfrac{12}{35}$

8) $\dfrac{1}{4} \div \dfrac{2}{5} = \dfrac{5}{8}$

Name:	Date: ....

Topic	**Adding Mixed Numbers**
Notes	Use the following steps for adding mixed numbers. ✓ Add whole numbers of the mixed numbers. ✓ Add the fractions of each mixed number. ✓ Find the Least Common Denominator (LCD) if necessary. ✓ Add whole numbers and fractions. ✓ Write your answer in lowest terms.
Example	***Add mixed numbers.*** $1\frac{1}{2} + 2\frac{2}{3} =$ Rewriting our equation with parts separated, $1 + \frac{1}{2} + 2 + \frac{2}{3}$ Add whole numbers: $1 + 2 = 3$ Add fractions: $\frac{1}{2} + \frac{2}{3} = \frac{3}{6} + \frac{4}{6} = \frac{7}{6} = 1\frac{1}{6}$, Now, combine the whole and fraction parts: $3 + 1 + \frac{1}{6} = 4\frac{1}{6}$
Your Turn!	1) $1\frac{1}{12} + 2\frac{3}{4} =$ 2) $3\frac{5}{8} + 1\frac{1}{4} =$ 3) $1\frac{1}{10} + 2\frac{2}{5} =$ 4) $2\frac{5}{6} + 2\frac{2}{9} =$ 5) $2\frac{2}{7} + 1\frac{2}{21} =$ 6) $1\frac{3}{8} + 3\frac{2}{3} =$ 7) $3\frac{1}{5} + 1\frac{2}{8} =$ 8) $3\frac{1}{2} + 2\frac{3}{7} =$

Name:	**Date:** ....

Topic	**Adding Mixed Numbers - Answers**
Notes	Use the following steps for adding mixed numbers. ✓ Add whole numbers of the mixed numbers. ✓ Add the fractions of each mixed number. ✓ Find the Least Common Denominator (LCD) if necessary. ✓ Add whole numbers and fractions. ✓ Write your answer in lowest terms.
Example	***Add mixed numbers.*** $1\frac{1}{2} + 2\frac{2}{3} =$ Rewriting our equation with parts separated, $1 + \frac{1}{2} + 2 + \frac{2}{3}$ Add whole numbers: $1 + 2 = 3$ Add fractions: $\frac{1}{2} + \frac{2}{3} = \frac{3}{6} + \frac{4}{6} = \frac{7}{6} = 1\frac{1}{6}$ Now, combine the whole and fraction parts: $3 + 1 + \frac{1}{6} = 4\frac{1}{6}$
Your Turn!	1) $1\frac{1}{12} + 2\frac{3}{4} = 3\frac{5}{6}$ 2) $3\frac{5}{8} + 1\frac{1}{4} = 4\frac{7}{8}$ 3) $1\frac{1}{10} + 2\frac{2}{5} = 3\frac{1}{2}$ 4) $2\frac{5}{6} + 2\frac{2}{9} = 5\frac{1}{18}$ 5) $2\frac{2}{7} + 1\frac{2}{21} = 3\frac{8}{21}$ 6) $1\frac{3}{8} + 3\frac{2}{3} = 5\frac{1}{24}$ 7) $3\frac{1}{5} + 1\frac{2}{8} = 4\frac{9}{20}$ 8) $3\frac{1}{2} + 2\frac{3}{7} = 5\frac{13}{14}$

Name:	Date: ....

Topic	**Subtracting Mixed Numbers**
Notes	Use the following steps for subtracting mixed numbers. ✓ Convert mixed numbers into improper fractions. $a\frac{c}{b} = \frac{ab+c}{b}$ ✓ Find equivalent fractions with the same denominator for unlike fractions (fractions with different denominators) ✓ Subtract the second fraction from the first one. ✓ Write your answer in lowest terms and convert it into a mixed number if the answer is an improper fraction.
Example	***Subtract.*** $5\frac{1}{2} - 2\frac{2}{3} =$ Convert mixed numbers into fractions: $5\frac{1}{2} = \frac{5\times2+1}{5} = \frac{11}{2}$ and $2\frac{2}{3} = \frac{2\times3+2}{4} = \frac{8}{3}$, these two fractions are "unlike" fractions. (they have different denominators). Find equivalent fractions with the same denominator. Use this formula: $\frac{a}{b} - \frac{c}{d} = \frac{ad-bc}{bd}$ $\frac{11}{2} - \frac{8}{3} = \frac{(11)(3)-(2)(8)}{2\times3} = \frac{33-16}{6} = \frac{17}{6}$, the answer is an improper fraction, convert it into a mixed number. $\frac{17}{6} = 2\frac{5}{6}$
Your Turn!	1) $2\frac{2}{5} - 1\frac{1}{3} =$ 2) $3\frac{5}{8} - 2\frac{1}{3} =$ 3) $6\frac{1}{4} - 1\frac{2}{7} =$ 4) $8\frac{2}{3} - 1\frac{1}{4} =$ 5) $8\frac{3}{4} - 1\frac{3}{8} =$ 6) $2\frac{3}{8} - 1\frac{2}{3} =$ 7) $13\frac{2}{7} - 1\frac{2}{21} =$ 8) $5\frac{1}{2} - 2\frac{3}{7} =$

Name: ....   **Date:** ....

Topic	**Subtracting Mixed Numbers - Answers**
Notes	Use the following steps for subtracting mixed numbers. ✓ Convert mixed numbers into improper fractions. $a\frac{c}{b} = \frac{ab+c}{b}$ ✓ Find equivalent fractions with the same denominator for unlike fractions (fractions with different denominators) ✓ Subtract the second fraction from the first one. ✓ Write your answer in lowest terms and convert it into a mixed number if the answer is an improper fraction.
Example	**Subtract.** $5\frac{1}{2} - 2\frac{2}{3} =$ Convert mixed numbers into fractions: $5\frac{1}{2} = \frac{5\times2+1}{5} = \frac{11}{2}$ and $2\frac{2}{3} = \frac{2\times3+2}{4} = \frac{8}{3}$, these two fractions are "unlike" fractions. (they have different denominators). Find equivalent fractions with the same denominator. Use this formula: $\frac{a}{b} - \frac{c}{d} = \frac{ad-bc}{bd}$ $\frac{11}{2} - \frac{8}{3} = \frac{(11)(3)-(2)(8)}{2\times3} = \frac{33-16}{6} = \frac{17}{6}$, the answer is an improper fraction, convert it into a mixed number. $\frac{17}{6} = 2\frac{5}{6}$
Your Turn!	1) $2\frac{2}{5} - 1\frac{1}{3} = 1\frac{1}{15}$ 2) $3\frac{5}{8} - 2\frac{1}{3} = 1\frac{7}{24}$ 3) $6\frac{1}{4} - 1\frac{2}{7} = 4\frac{27}{28}$ 4) $8\frac{2}{3} - 1\frac{1}{4} = 7\frac{5}{12}$ 5) $8\frac{3}{4} - 1\frac{3}{8} = 7\frac{3}{8}$ 6) $2\frac{3}{8} - 1\frac{2}{3} = \frac{17}{24}$ 7) $13\frac{2}{7} - 1\frac{2}{21} = 12\frac{4}{21}$ 8) $5\frac{1}{2} - 2\frac{3}{7} = 3\frac{1}{14}$

Name:	Date: ....

Topic	**Multiplying Mixed Numbers**
Notes	✓ Convert the mixed numbers into fractions. $a\frac{c}{b} = a + \frac{c}{b} = \frac{ab+c}{b}$ ✓ Multiply fractions and simplify if necessary. $\frac{a}{b} \times \frac{c}{d} = \frac{a \times c}{b \times d}$ ✓ If the answer is an improper fraction (numerator is bigger than denominator), convert it into a mixed number.
Example	***Multiply*** $2\frac{1}{4} \times 3\frac{1}{2}$ Convert mixed numbers into fractions: $2\frac{1}{4} = \frac{2 \times 4 + 1}{4} = \frac{9}{4}$ and $3\frac{1}{2} = \frac{3 \times 2 + 1}{2} = \frac{7}{2}$ Multiply two fractions: $\frac{9}{4} \times \frac{7}{2} = \frac{9 \times 7}{4 \times 2} = \frac{63}{8}$ The answer is an improper fraction. Convert it into a mixed number: $$\frac{63}{8} = 7\frac{7}{8}$$
Your Turn!	1) $5\frac{2}{3} \times 2\frac{2}{9} =$ 2) $4\frac{1}{6} \times 5\frac{3}{7} =$ 3) $3\frac{1}{3} \times 3\frac{3}{4} =$ 4) $2\frac{2}{9} \times 6\frac{1}{3} =$ 5) $2\frac{2}{7} \times 4\frac{3}{5} =$ 6) $1\frac{4}{7} \times 9\frac{1}{2} =$ 7) $4\frac{1}{8} \times 3\frac{2}{3} =$ 8) $6\frac{2}{3} \times 1\frac{1}{4} =$

Name:	**Date:** ....

Topic	**Multiplying Mixed Numbers - Answers**
Notes	✓ Convert the mixed numbers into fractions. $a\dfrac{c}{b} = a + \dfrac{c}{b} = \dfrac{ab+c}{b}$ ✓ Multiply fractions and simplify if necessary. $\dfrac{a}{b} \times \dfrac{c}{d} = \dfrac{a \times c}{b \times d}$ ✓ If the answer is an improper fraction (numerator is bigger than denominator), convert it into a mixed number.
Example	**Multiply** $2\dfrac{1}{4} \times 3\dfrac{1}{2}$ Convert mixed numbers into fractions: $2\dfrac{1}{4} = \dfrac{2 \times 4 + 1}{4} = \dfrac{9}{4}$ and $3\dfrac{1}{2} = \dfrac{3 \times 2 + 1}{2} = \dfrac{7}{2}$ Multiply two fractions: $\dfrac{9}{4} \times \dfrac{7}{2} = \dfrac{9 \times 7}{4 \times 2} = \dfrac{63}{8}$ The answer is an improper fraction. Convert it into a mixed number: $$\dfrac{63}{8} = 7\dfrac{7}{8}$$
Your Turn!	1) $5\dfrac{2}{3} \times 2\dfrac{2}{9} = 12\dfrac{16}{27}$ 2) $4\dfrac{1}{6} \times 5\dfrac{3}{7} = 22\dfrac{13}{21}$ 3) $3\dfrac{1}{3} \times 3\dfrac{3}{4} = 12\dfrac{1}{2}$ 4) $2\dfrac{2}{9} \times 6\dfrac{1}{3} = 14\dfrac{2}{27}$ 5) $2\dfrac{2}{7} \times 4\dfrac{3}{5} = 10\dfrac{18}{35}$ 6) $1\dfrac{4}{7} \times 9\dfrac{1}{2} = 14\dfrac{13}{14}$ 7) $4\dfrac{1}{8} \times 3\dfrac{2}{3} = 15\dfrac{1}{8}$ 8) $6\dfrac{2}{3} \times 1\dfrac{1}{4} = 8\dfrac{1}{3}$

| Name: .... | Date: .... |

Topic	**Dividing Mixed Numbers**
Notes	✓ Convert the mixed numbers into improper fractions. $$a\frac{c}{b} = a + \frac{c}{b} = \frac{ab+c}{b}$$ ✓ Divide fractions and simplify if necessary.
Example	*Solve.* $2\frac{1}{3} \div 1\frac{1}{4} =$ Converting mixed numbers to fractions: $2\frac{1}{3} \div 1\frac{1}{4} = \frac{7}{3} \div \frac{5}{4}$ Keep, Change, Flip: $\frac{7}{3} \div \frac{5}{4} = \frac{7}{3} \times \frac{4}{5} = \frac{7\times4}{3\times5} = \frac{28}{15} = 1\frac{13}{15}$

Your Turn!	1) $3\frac{2}{7} \div 2\frac{1}{4} =$	2) $4\frac{2}{9} \div 1\frac{5}{6} =$
	3) $4\frac{2}{3} \div 3\frac{2}{5} =$	4) $5\frac{4}{5} \div 4\frac{3}{4} =$
	5) $1\frac{8}{9} \div 2\frac{3}{7} =$	6) $3\frac{3}{8} \div 2\frac{2}{5} =$
	7) $4\frac{1}{5} \div 3\frac{1}{9} =$	8) $4\frac{2}{3} \div 1\frac{8}{9} =$
	9) $5\frac{2}{3} \div 3\frac{3}{7} =$	10) $7\frac{1}{2} \div 5\frac{1}{3} =$

Name:	Date: ....

Topic	Dividing Mixed Numbers- Answers	
Notes	✓ Convert the mixed numbers into improper fractions. $$a\frac{c}{b} = a + \frac{c}{b} = \frac{ab + c}{b}$$ ✓ Divide fractions and simplify if necessary.	
Example	**Solve.** $2\frac{1}{3} \div 1\frac{1}{4} =$ Converting mixed numbers to fractions: $2\frac{1}{3} \div 1\frac{1}{4} = \frac{7}{3} \div \frac{5}{4}$ Keep, Change, Flip: $\frac{7}{3} \div \frac{5}{4} = \frac{7}{3} \times \frac{4}{5} = \frac{7\times4}{3\times5} = \frac{28}{15} = 1\frac{13}{15}$	
Your Turn!	1) $3\frac{2}{7} \div 2\frac{1}{4} = 1\frac{29}{63}$	2) $4\frac{2}{9} \div 1\frac{5}{6} = 2\frac{10}{33}$
	3) $4\frac{2}{3} \div 3\frac{2}{5} = 1\frac{19}{51}$	4) $5\frac{4}{5} \div 4\frac{3}{4} = 1\frac{21}{95}$
	5) $1\frac{8}{9} \div 2\frac{3}{7} = \frac{7}{9}$	6) $3\frac{3}{8} \div 2\frac{2}{5} = 1\frac{13}{32}$
	7) $4\frac{1}{5} \div 3\frac{1}{9} = 1\frac{7}{20}$	8) $4\frac{2}{3} \div 1\frac{8}{9} = 2\frac{8}{17}$
	9) $5\frac{2}{3} \div 3\frac{3}{7} = 1\frac{47}{72}$	10) $7\frac{1}{2} \div 5\frac{1}{3} = 1\frac{13}{32}$

Name:	Date: ....

Topic	**Comparing Decimals**
Notes	Decimals: is a fraction written in a special form. For example, instead of writing $\frac{1}{2}$ you can write **0.5**. For comparing decimals: ✓ Compare each digit of two decimals in the same place value. ✓ Start from left. Compare hundreds, tens, ones, tenth, hundredth, etc. ✓ To compare numbers, use these symbols: - Equal to $=$, Less than $<$, Greater than $>$ Greater than or equal $\geq$, Less than or equal $\leq$
Examples	**Compare 0.40 and 0.04.** 0.40 *is greater than* 0.04, because the tenth place of 0.40 is 4, but the tenth place of 0.04 is zero. Then: $0.40 > 0.04$ **Compare 0.0912 and 0.912.** 0.912 *is greater than* 0.0912, because the tenth place of 0.912 is 9, but the tenth place of 0.0912 is zero. Then: $0.0912 < 0.912$
Your Turn!	1) 0.91 ☐ 0.95 2) 1.79 ☐ 1.80 3) 19.1 ☐ 19.09 4) 2.45 ☐ 2.089 5) 1.258 ☐ 12.58 6) 0.89 ☐ 0.890 7) 3.871 ☐ 2.998 8) 0.567 ☐ 0.756

Name:	Date: ....

Topic	**Comparing Decimals - Answers**	
Notes	Decimals: is a fraction written in a special form. For example, instead of writing $\frac{1}{2}$ you can write **0.5**. For comparing decimals: ✓ Compare each digit of two decimals in the same place value. ✓ Start from left. Compare hundreds, tens, ones, tenth, hundredth, etc. ✓ To compare numbers, use these symbols: - Equal to $=$, Less than $<$, Greater than $>$ Greater than or equal $\geq$, Less than or equal $\leq$	
Examples	**Compare 0.40 and 0.04.** 0.40 *is greater than* 0.04, because the tenth place of 0.40 is 4, but the tenth place of 0.04 is zero. Then: $0.40 > 0.04$ **Compare 0.0912 and 0.912.** 0.912 *is greater than* 0.0912, because the tenth place of 0.912 is 9, but the tenth place of 0.0912 is zero. Then: $0.0912 < 0.912$	
Your Turn!	1) $0.91 < 0.95$	2) $1.78 < 1.80$
	3) $19.1 > 19.09$	4) $2.45 > 2.089$
	5) $1.258 < 12.58$	6) $0.89 = 0.890$
	7) $3.387 > 2.998$	8) $0.567 < 0.756$

Name:	Date: ....

Topic	**Rounding Decimals**
Notes	✓ We can round decimals to a certain accuracy or number of decimal places. ✓ Let's review place values: For example: **35.4817** 3: tens 5: ones 4: tenths 8: hundredths 1: thousandths 7:tens thousandths ✓ To round a decimal, find the place value you'll round to. ✓ Find the digit to the right of the place value you're rounding to. If it is 5 or bigger, add 1 to the place value you're rounding to and remove all digits on its right side. If the digit to the right of the place value is less than 5, keep the place value and remove all digits on the right.
Example	*Round 12.8365 to the hundredth place value.* First look at the next place value to the right, (thousandths). It's 6 and it is greater than 5. Thus add 1 to the digit in the hundredth place. It is 3. $\rightarrow$ $3 + 1 = 4$, then, the answer is 12.84
Your Turn!	*Round each number to the underlined place value.* 1) 32.5$\underline{4}$8 = 2) 2.3$\underline{2}$6 = 3) 55.$\underline{4}$23 = 4) 2$\underline{5}$.62 = 5) 11.$\underline{2}$65 = 6) 33.5$\underline{0}$5 = 7) 3.5$\underline{8}$9 = 8) 8.0$\underline{1}$9 =

Name:	Date: ....

Topic	Rounding Decimals - Answers
Notes	✓ We can round decimals to a certain accuracy or number of decimal places. ✓ Let's review place values: For example: $$35.4817$$ 3: tens 5: ones 4: tenths 8: hundredths 1: thousandths 7:tens thousandths ✓ To round a decimal, find the place value you'll round to. ✓ Find the digit to the right of the place value you're rounding to. If it is 5 or bigger, add 1 to the place value you're rounding to and remove all digits on its right side. If the digit to the right of the place value is less than 5, keep the place value and remove all digits on the right.
Example	***Round 12.8365 to the hundredth place value.*** First look at the next place value to the right, (thousandths). It's 6 and it is greater than 5. Thus add 1 to the digit in the hundredth place. It is 3. → $3 + 1 = 4$, then, the answer is 12.84
Your Turn!	***Round each number to the underlined place value.*** 1) $32.5\underline{4}8 = 32.55$ | 2) $2.3\underline{2}6 = 2.33$ 3) $55.\underline{4}23 = 55.4$ | 4) $2\underline{5}.62 = 26$ 5) $11.\underline{2}65 = 11.3$ | 6) $33.5\underline{0}5 = 33.51$ 7) $3.5\underline{8}9 = 3.59$ | 8) $8.0\underline{1}9 = 8.02$

Name:	Date: ....

Topic	**Adding and Subtracting Decimals**
Notes	✓ Line up the numbers. ✓ Add zeros to have same number of digits for both numbers if necessary. ✓ Add or subtract using column addition or subtraction.
Examples	***Add***. $2.6 + 5.33 =$ First line up the numbers: $\begin{array}{r} 2.6 \\ +\,5.33 \\ \hline \end{array}$ →Add zeros to have same number of digits for both numbers. $\begin{array}{r} 2.60 \\ +\,5.33 \\ \hline \end{array}$ → Start with the hundredths place. $0 + 3 = 2$, $\begin{array}{r} 2.60 \\ +\,5.33 \\ \hline 3 \end{array}$ → Continue with tenths place. $6 + 3 = 9$, $\begin{array}{r} 2.60 \\ +\,5.33 \\ \hline .93 \end{array}$ → Add the ones place. $2 + 5 = 7$, $\begin{array}{r} 2.60 \\ +\,5.33 \\ \hline 7.93 \end{array}$ ***Subtract***. $4.79 - 3.13 =$ $\begin{array}{r} 4.79 \\ -\,3.13 \\ \hline \end{array}$ Start with the hundredths place. $9 - 3 = 6$, $\begin{array}{r} 4.79 \\ -\,3.13 \\ \hline 6 \end{array}$, continue with tenths place. $7 - 1 = 6$, $\begin{array}{r} 4.79 \\ -\,3.13 \\ \hline .66 \end{array}$, subtract the ones place. $4 - 3 = 1$, $\begin{array}{r} 4.79 \\ -\,3.13 \\ \hline 1.66 \end{array}$
Your Turn!	1) $48.13 + 20.15 =$ 2) $78.14 - 65.19 =$ 3) $38.19 + 24.18 =$ 4) $57.26 - 43.54 =$ 5) $27.89 + 46.13 =$ 6) $49.65 - 32.78 =$

| Name: .... | Date: .... |

Topic	**Adding and Subtracting Decimals - Answers**	
Notes	✓ Line up the numbers. ✓ Add zeros to have same number of digits for both numbers if necessary. ✓ Add or subtract using column addition or subtraction.	
Examples	***Add.*** $2.6 + 5.33 =$ First line up the numbers: $\begin{array}{r} 2.6 \\ +\,5.33 \\ \hline \end{array}$ →Add zeros to have same number of digits for both numbers. $\begin{array}{r} 2.60 \\ +\,5.33 \\ \hline \end{array}$ → Start with the hundredths place. $0 + 3 = 2$, $\begin{array}{r} 2.60 \\ +\,5.33 \\ \hline 3 \end{array}$ → Continue with tenths place. $6 + 3 = 9$, $\begin{array}{r} 2.60 \\ +\,5.33 \\ \hline .93 \end{array}$ → Add the ones place. $2 + 5 = 7$, $\begin{array}{r} 2.60 \\ +\,5.33 \\ \hline 7.93 \end{array}$ ***Subtract.*** $4.79 - 3.13 =$ $\begin{array}{r} 4.79 \\ -\,3.13 \\ \hline \end{array}$ Start with the hundredths place. $9 - 3 = 6$, $\begin{array}{r} 4.79 \\ -\,3.13 \\ \hline 6 \end{array}$, continue with tenths place. $7 - 1 = 6$, $\begin{array}{r} 4.79 \\ -\,3.13 \\ \hline .66 \end{array}$, subtract the ones place. $4 - 3 = 1$, $\begin{array}{r} 4.79 \\ -\,3.13 \\ \hline 1.66 \end{array}$	
Your Turn!	1) $48.13 + 20.15 = 68.28$	2) $78.14 - 65.19 = 12.95$
	3) $38.19 + 24.18 = 62.37$	4) $57.26 - 43.54 = 13.72$
	5) $27.89 + 46.13 = 74.02$	6) $49.65 - 32.78 = 16.87$

Name:	Date: ....

Topic	**Multiplying and Dividing Decimals**
Notes	For Multiplication: ✓ Ignore the decimal point and set up and multiply the numbers as you do with whole numbers. ✓ Count the total number of decimal places in both factors. ✓ Place the decimal point in the product. For Division: ✓ If the divisor is not a whole number, move decimal point to right to make it a whole number. Do the same for dividend. ✓ Divide similar to whole numbers.
Examples	**Find the product.** $1.2 \times 2.3 =$ Set up and multiply the numbers as you do with whole numbers. Line up the numbers: $\begin{array}{r} 12 \\ \times\,23 \\ \hline \end{array} \rightarrow$ Multiply: $\begin{array}{r} 12 \\ \times\,23 \\ \hline 276 \end{array} \rightarrow$ Count the total number of decimal places in both of the factors. There are two decimal digits. Then: $1.2 \times 2.3 = 2.76$ **Find the quotient.** $5.6 \div 0.8 =$ The divisor is not a whole number. Multiply it by 10 to get 8. $\rightarrow 0.8 \times 10 = 8$ Do the same for the dividend to get 56 $\rightarrow 5.6 \times 10 = 56$ Now, divide: $56 \div 8 = 7$. The answer is 7.
Your Turn!	1) $1.13 \times 0.7 =$ 2) $48.8 \div 8 =$ 3) $0.9 \times 0.68 =$ 4) $66.8 \div 0.2 =$ 5) $0.18 \times 0.5 =$ 6) $37.2 \div 100 =$

Name:	**Date:** ....

Topic	**Multiplying and Dividing Decimals - Answers**
Notes	For Multiplication: ✓ Ignore the decimal point and set up and multiply the numbers as you do with whole numbers. ✓ Count the total number of decimal places in both factors. ✓ Place the decimal point in the product. For Division: ✓ If the divisor is not a whole number, move decimal point to right to make it a whole number. Do the same for dividend. ✓ Divide similar to whole numbers.
Examples	***Find the product.*** $1.2 \times 2.3 =$ Set up and multiply the numbers as you do with whole numbers. Line up the numbers: $\begin{array}{r} 12 \\ \times 23 \\ \hline \end{array}$ → Multiply: $\begin{array}{r} 12 \\ \times 23 \\ \hline 276 \end{array}$ → Count the total number of decimal places in both of the factors. There are two decimal digits. Then: $1.2 \times 2.3 = 2.76$ ***Find the quotient.*** $5.6 \div 0.8 =$ The divisor is not a whole number. Multiply it by 10 to get 8. → $0.8 \times 10 = 8$ Do the same for the dividend to get 56 → $5.6 \times 10 = 56$ Now, divide: $56 \div 8 = 7$. The answer is 7.

Your Turn!	1) $1.13 \times 0.7 = 0.791$	2) $48.8 \div 8 = 6.1$
	3) $0.9 \times 0.68 = 0.612$	4) $66.8 \div 0.2 = 334$
	5) $0.18 \times 0.5 = 0.09$	6) $37.2 \div 100 = 0.372$

Name:	Date: ....

Topic	**Adding and Subtracting Integers**
Notes	✓ Integers include: zero, counting numbers, and the negative of the counting numbers. $\{\dots, -3, -2, -1, 0, 1, 2, 3, \dots\}$ ✓ Add a positive integer by moving to the right on the number line. ✓ Add a negative integer by moving to the left on the number line. Subtract an integer by adding its opposite.
Examples	**_Solve._** $(4) - (-8) =$ Keep the first number and convert the sign of the second number to its opposite. (change subtraction into addition. Then: $(4) + 8 = 12$ **_Solve._** $42 + (12 - 26) =$ First subtract the numbers in brackets, $12 - 26 = -14$ Then: $42 + (-14) = \rightarrow$ change addition into subtraction: $42 - 14 = 28$
Your Turn!	1) $-(15) + 12 =$ 2) $(-2) + (-10) + 18 =$ 3) $(-13) + 7 =$ 4) $3 - (-7) + 14 =$ 5) $(-7) + (-8) =$ 6) $16 - (-4 + 8) =$ 7) $4 + (-15) + 2 =$ 8) $-(22) - (-4) + 8 =$

Name:	Date: ....

Topic	**Adding and Subtracting Integers - Answers**
Notes	✓ Integers include: zero, counting numbers, and the negative of the counting numbers. $\{... , -3, -2, -1, 0, 1, 2, 3, ...\}$ ✓ Add a positive integer by moving to the right on the number line. ✓ Add a negative integer by moving to the left on the number line. Subtract an integer by adding its opposite.
Examples	***Solve.*** $(4) - (-8) =$ Keep the first number and convert the sign of the second number to its opposite. (change subtraction into addition. Then: $(4) + 8 = 12$ ***Solve.*** $42 + (12 - 26) =$ First subtract the numbers in brackets, $12 - 26 = -14$ Then: $42 + (-14) = \;\rightarrow$ change addition into subtraction: $42 - 14 = 28$

Your Turn!	1) $-(15) + 12 = -3$	2) $(-2) + (-10) + 18 = 6$
	3) $(-13) + 7 = -6$	4) $3 - (-7) + 14 = 24$
	5) $(-7) + (-8) = -15$	6) $16 - (-4 + 8) = 12$
	7) $4 + (-15) + 2 = -9$	8) $(-22) - (-4) + 8 = -10$

Name:	Date: ....

Topic	**Multiplying and Dividing Integers**
Notes	Use following rules for multiplying and dividing integers: ✓ (negative) × (negative) = positive ✓ (negative) ÷ (negative) = positive ✓ (negative) × (positive) = negative ✓ (negative) ÷ (positive) = negative ✓ (positive) × (positive) = positive ✓ (positive) ÷ (negative) = negative
Examples	***Solve.*** $2 \times (14 - 17) =$ First subtract the numbers in brackets, $14 - 17 = -3 \rightarrow (2) \times (-3) =$ Now use this rule: (positive) × (negative) = negative $(2) \times (-3) = -6$ ***Solve.*** $(-7) + (-36 \div 4) =$ First divide -36 by 4, the numbers in brackets, using this rule: (negative) ÷ (positive) = negative Then: $-36 \div 4 = -9$. Now, add -7 and -9: $(-7) + (-9) = -7 - 9 = -16$
Your Turn!	1) $(-7) \times 6 =$ 2) $(-63) \div (-7) =$ 3) $(-11) \times (-3) =$ 4) $81 \div (-9) =$ 5) $(15 - 12) \times (-7) =$ 6) $(-12) \div (3) =$ 7) $4 \times (-9) =$ 8) $(8) \div (-2) =$

Name:	Date: ....

Topic	**Multiplying and Dividing Integers - Answers**	
Notes	Use following rules for multiplying and dividing integers: ✓ (negative) × (negative) = positive ✓ (negative) ÷ (negative) = positive ✓ (negative) × (positive) = negative ✓ (negative) ÷ (positive) = negative ✓ (positive) × (positive) = positive ✓ (positive) ÷ (negative) = negative	
Examples	***Solve.*** $2 \times (14 - 17) =$ First subtract the numbers in brackets, $14 - 17 = -3 \rightarrow (2) \times (-3) =$ Now use this rule: (positive) × (negative) = negative $(2) \times (-3) = -6$ ***Solve.*** $(-7) + (-36 \div 4) =$ First divide -36 by 4 , the numbers in brackets, using this rule: (negative) ÷ (positive) = negative Then: $-36 \div 4 = -9$. Now, add -7 and -9: $\qquad (-7) + (-9) = -7 - 9 = -16$	
Your Turn!	1) $(-7) \times 6 = -42$	2) $(-63) \div (-7) = 9$
	3) $(-11) \times (-3) = 33$	4) $81 \div (-9) = -9$
	5) $(15 - 12) \times (-7) = -21$	6) $(-12) \div (3) = -4$
	7) $4 \times (-9) = -36$	8) $(8) \div (-2) = -4$

Name:	Date: ....

Topic	**Order of Operation**
Notes	When there is more than one math operation, use PEMDAS: (to memorize this rule, remember the phrase "Please Excuse My Dear Aunt Sally") ✓ Parentheses ✓ Exponents ✓ Multiplication and Division (from left to right) ✓ Addition and Subtraction (from left to right)
Examples	**_Calculate._** $(18 - 26) \div (2^4 \div 4) =$ First simplify inside parentheses: $(-8) \div (16 \div 4) = (-8) \div (4)$ Then: $(-8) \div (4) = -2$ **_Solve._** $(-5 \times 7) - (18 - 3^2) =$ First calculate within parentheses: $(-5 \times 7) - (18 - 3^2) = (-35) - (18 - 9)$ Then: $(-35) - (18 - 9) = -35 - 9 = -44$
Your Turn!	1) $(11 \times 4) \div (5 + 6) =$ 2) $(30 \div 5) + (17 - 8) =$ 3) $(-9) + (5 \times 6) + 14 =$ 4) $(-10 \times 5) \div (2^2 + 1) =$ 5) $[-16(32 \div 2^3)] \div 8 =$ 6) $(-7) + (72 \div 3^2) + 12 =$ 7) $[16(32 \div 2^3)] - 4^2 =$ 8) $4^3 + (-5 \times 2^5) + 5 =$

Name:	Date: ....

Topic	**Order of Operation - Answers**
Notes	When there is more than one math operation, use PEMDAS: (to memorize this rule, remember the phrase "Please Excuse My Dear Aunt Sally") ✓ Parentheses ✓ Exponents ✓ Multiplication and Division (from left to right) ✓ Addition and Subtraction (from left to right)
Examples	***Calculate.*** $(18 - 26) \div (2^4 \div 4) =$ First simplify inside parentheses: $(-8) \div (16 \div 4) = (-8) \div (4)$ Then: $(-8) \div (4) = -2$ ***Solve.*** $(-5 \times 7) - (18 - 3^2) =$ First calculate within parentheses: $(-5 \times 7) - (18 - 3^2) = (-35) - (18 - 9)$ Then: $(-35) - (18 - 9) = -35 - 9 = -44$

Your Turn!	1) $(11 \times 4) \div (5 + 6) = 4$	2) $(30 \div 5) + (17 - 8) = 15$
	3) $(-9) + (5 \times 6) + 14 =$ 35	4) $(-10 \times 5) \div (2^2 + 1) = -10$
	5) $[-16(32 \div 2^3)] \div 8 =$ -8	6) $(-7) + (72 \div 3^2) + 12 = 13$
	7) $[16(32 \div 2^3)] - 4^2 =$ 48	8) $4^3 + (-5 \times 2^5) + 5 = -91$

Name:	Date: ....

Topic	**Integers and Absolute Value**
Notes	✓ The absolute value of a number is its distance from zero, in either direction, on the number line. For example, the distance of 9 and -9 from zero on number line is 9. ✓ Absolute value is symbolized by vertical bars, as in $|x|$.
Example	*Calculate.* $|8 - 5| \times |12 - 16| =$ First calculate $|8 - 5|$, $\rightarrow |8 - 5| = |3|$, the absolute value of 3 is 3, $|3| = 3$ $8 \times |12 - 16| =$ Now calculate $|12 - 16|$, $\rightarrow |12 - 16| = |-4|$, the absolute value of -4 is 4, $|-4| = 4$. Then: $3 \times 4 = 12$

Your Turn!	
1) $11 - |4 - 13| =$	2) $14 - |12 - 19| - |9| =$
3) $|21| - \dfrac{|-25|}{5} =$	4) $|30| + \dfrac{|-49|}{7} =$
5) $\dfrac{|7 \times -8|}{4} \times \dfrac{|-12|}{2} =$	6) $\dfrac{|10 \times -6|}{5} \times |-9| =$
7) $\dfrac{|-20|}{5} \times \dfrac{|-36|}{6} =$	8) $|-30 + 6| \times \dfrac{|-9 \times 4|}{12} =$

Name:	Date: ....

Topic	Integers and Absolute Value - Answers
Notes	✓ The absolute value of a number is its distance from zero, in either direction, on the number line. For example, the distance of 9 and -9 from zero on number line is 9. ✓ Absolute value is symbolized by vertical bars, as in $\|x\|$.
Example	*Calculate.* $\|8-5\| \times \|12-16\| =$ First calculate $\|8-5\|$, $\rightarrow \|8-5\| = \|3\|$, the absolute value of 3 is 3, $\|3\| = 3$ $8 \times \|12-16\| =$ Now calculate $\|12-16\|$, $\rightarrow \|12-16\| = \|-4\|$, the absolute value of -4 is 4, $\|-4\| = 4$. Then: $3 \times 4 = 12$

Your Turn!	1) $11 - \|4 - 13\| = 2$	2) $14 - \|12 - 19\| - \|9\| = -2$
	3) $\|21\| - \dfrac{\|-25\|}{5} = 16$	4) $\|30\| + \dfrac{\|-49\|}{7} = 37$
	5) $\dfrac{\|7 \times -8\|}{4} \times \dfrac{\|-12\|}{2} = 84$	6) $\dfrac{\|10 \times -6\|}{5} \times \|-9\| = 108$
	7) $\dfrac{\|-20\|}{5} \times \dfrac{\|-36\|}{6} = 24$	8) $\|-30 + 6\| \times \dfrac{\|-9 \times 4\|}{12} = 72$

| Name: .... | Date: .... |

Topic	**Simplifying Ratios**
Notes	✓ Ratios are used to make comparisons between two numbers. ✓ Ratios can be written as a fraction, using the word "to", or with a colon. ✓ You can calculate equivalent ratios by multiplying or dividing both sides of the ratio by the same number.
Examples	*Simplify.* $18 : 63 =$ Both numbers 18 and 63 are divisible by $9 \Rightarrow 18 \div 9 = 2$, $63 \div 9 = 7$, Then: $18 : 63 = 2 : 7$ *Simplify.* $\dfrac{25}{45} =$ Both numbers 25 and 45 are divisible by 5, $\Rightarrow 25 \div 5 = 5$, $45 \div 5 = 9$, Then: $\dfrac{25}{45} = \dfrac{5}{9}$
Your Turn!	1) $\dfrac{4}{32} = -$ 2) $\dfrac{25}{80} = -$ 3) $\dfrac{15}{35} = -$ 4) $\dfrac{42}{54} = -$ 5) $\dfrac{12}{36} = -$ 6) $\dfrac{30}{80} = -$ 7) $\dfrac{18}{24} = -$ 8) $\dfrac{60}{108} = -$

Name:	Date: ....

Topic	**Simplifying Ratios - Answers**
Notes	✓ Ratios are used to make comparisons between two numbers. ✓ Ratios can be written as a fraction, using the word "to", or with a colon. ✓ You can calculate equivalent ratios by multiplying or dividing both sides of the ratio by the same number.
Examples	*Simplify.* $18:63 =$ Both numbers 18 and 63 are divisible by $9 \Rightarrow 18 \div 9 = 2, 63 \div 9 = 7,$ Then: $18:63 = 2:7$ *Simplify.* $\frac{25}{45} =$ Both numbers 25 and 45 are divisible by $5, \Rightarrow 25 \div 5 = 5, 45 \div 5 = 9,$ Then: $\frac{25}{45} = \frac{5}{9}$

Your Turn!	1) $\frac{4}{32} = \frac{1}{8}$	2) $\frac{25}{80} = \frac{5}{16}$
	3) $\frac{15}{35} = \frac{3}{7}$	4) $\frac{42}{54} = \frac{7}{9}$
	5) $\frac{12}{36} = \frac{1}{3}$	6) $\frac{30}{80} = \frac{3}{8}$ 7)
	8) $\frac{18}{24} = \frac{3}{4}$	9) $\frac{60}{108} = \frac{5}{9}$

| Name: .... | Date: .... |

Topic	**Proportional Ratios**
Notes	✓ Two ratios are proportional if they represent the same relationship. ✓ A proportion means that two ratios are equal. It can be written in two ways: $\dfrac{a}{b} = \dfrac{c}{d}$ $\qquad a : b = c : d$
Example	***Solve this proportion for*** x. $\dfrac{5}{8} = \dfrac{35}{x}$ Use cross multiplication: $\dfrac{5}{8} = \dfrac{35}{x} \Rightarrow 5 \times x = 8 \times 35 \Rightarrow 5x = 280$ Divide to find x: $\quad x = \dfrac{280}{5} \Rightarrow x = 56$
Your Turn!	1) $\dfrac{1}{9} = \dfrac{8}{x} \Rightarrow x = $ ____ 3) $\dfrac{3}{11} = \dfrac{6}{x} \Rightarrow x = $ ____ 5) $\dfrac{9}{12} = \dfrac{27}{x} \Rightarrow x = $ ____ 7) $\dfrac{7}{15} = \dfrac{49}{x} \Rightarrow x = $ ____ 2) $\dfrac{5}{8} = \dfrac{25}{x} \Rightarrow x = $ ____ 4) $\dfrac{12}{20} = \dfrac{x}{200} \Rightarrow x = $ ____ 6) $\dfrac{14}{16} = \dfrac{x}{80} \Rightarrow x = $ ____ 8) $\dfrac{8}{19} = \dfrac{32}{x} \Rightarrow x = $ ____

| Name: .... | Date: .... |

Topic	**Proportional Ratios - Answers**
Notes	✓ Two ratios are proportional if they represent the same relationship. ✓ A proportion means that two ratios are equal. It can be written in two ways: $\frac{a}{b} = \frac{c}{d}$ $\qquad\qquad a : b = c : d$
Example	**Solve this proportion for** x. $\frac{5}{8} = \frac{35}{x}$ Use cross multiplication: $\frac{5}{8} = \frac{35}{x} \Rightarrow 5 \times x = 8 \times 35 \Rightarrow 5x = 280$ Divide to find x: $\quad x = \frac{280}{5} \Rightarrow x = 56$

Your Turn!	1) $\frac{1}{9} = \frac{8}{x} \Rightarrow x = 72$	2) $\frac{5}{8} = \frac{25}{x} \Rightarrow x = 40$
	3) $\frac{3}{11} = \frac{6}{x} \Rightarrow x = 22$	4) $\frac{12}{20} = \frac{x}{200} \Rightarrow x = 120$
	5) $\frac{9}{12} = \frac{27}{x} \Rightarrow x = 36$	6) $\frac{14}{16} = \frac{x}{80} \Rightarrow x = 70$
	7) $\frac{7}{15} = \frac{49}{x} \Rightarrow x = 105$	8) $\frac{8}{19} = \frac{32}{x} \Rightarrow x = 76$

| **Name:** .... | **Date:** .... |

Topic	**Create Proportion**
Notes	✓ To create a proportion, simply find (or create) two equal fractions. ✓ Use cross products to solve proportions or to test whether two ratios are equal and form a proportion. $\frac{a}{b} = \frac{c}{d} \Rightarrow a \times d = c \times b$
Example	*State if this pair of ratios form a proportion.* $\frac{2}{3}$ *and* $\frac{12}{30}$ Use cross multiplication: $\frac{2}{3} = \frac{12}{30} \rightarrow 2 \times 30 = 12 \times 3 \rightarrow 60 = 36$, which is not correct. Therefore, this pair of ratios doesn't form a proportion.
Your Turn!	*State if each pair of ratios form a proportion.* 1) $\frac{4}{8}$ *and* $\frac{24}{48}$ 2) $\frac{5}{15}$ *and* $\frac{10}{20}$ 3) $\frac{3}{11}$ *and* $\frac{9}{33}$ 4) $\frac{7}{10}$ *and* $\frac{14}{20}$ 5) $\frac{7}{9}$ *and* $\frac{48}{81}$ 6) $\frac{6}{8}$ *and* $\frac{12}{14}$ 7) $\frac{2}{10}$ *and* $\frac{6}{30}$ 8) $\frac{9}{12}$ *and* $\frac{18}{24}$ 9) Solve. Five pencils costs \$0.65. How many pencils can you buy for \$2.60? _____

Name: ....

Date: ....

Topic	Create Proportion
Notes	✓ To create a proportion, simply find (or create) two equal fractions. ✓ Use cross products to solve proportions or to test whether two ratios are equal and form a proportion. $\frac{a}{b} = \frac{c}{d} \Rightarrow a \times d = c \times b$
Example	***State if this pair of ratios form a proportion.*** $\frac{2}{3} \ and \ \frac{12}{30}$ Use cross multiplication: $\frac{2}{3} = \frac{12}{30} \to 2 \times 30 = 12 \times 3 \to 60 = 36$, which is not correct. Therefore, this pair of ratios doesn't form a proportion.
Your Turn!	***State if each pair of ratios form a proportion.*** 1) $\frac{4}{8} \ and \ \frac{24}{48}, Yes$ 2) $\frac{5}{15} \ and \ \frac{10}{20}, No$ 3) $\frac{3}{11} \ and \ \frac{9}{33}, Yes$ 4) $\frac{7}{10} \ and \ \frac{14}{20}, Yes$ 5) $\frac{7}{9} \ and \ \frac{48}{81}, No$ 6) $\frac{6}{8} \ and \ \frac{12}{14}, No$ 7) $\frac{2}{10} \ and \ \frac{6}{30}, Yes$ 8) $\frac{9}{12} \ and \ \frac{18}{24}, Yes$ 9) Solve. Five pencils costs $0.65. How many pencils can you buy for $2.60? **20 pencils**

Name:	Date: ....

Topic	**Similarity and Ratios**
Notes	✓ Two figures are similar if they have the same shape. ✓ Two or more figures are similar if the corresponding angles are equal, and the corresponding sides are in proportion.
Example	*Following triangles are similar. What is the value of unknown side?* **Solution:** Find the corresponding sides and write a proportion: $\frac{4}{12} = \frac{x}{9}$. Now, use cross product to solve for x: $\frac{4}{12} = \frac{x}{9} \rightarrow 4 \times 9 = 12 \times x \rightarrow 36 = 12x$. Divide both sides by 12. Then: $5x = 40 \rightarrow \frac{36}{12} = \frac{12x}{12} \rightarrow x = 3$. The missing side is 3.
Your Turn!	1) 2) 3) 4) 5) 6)

Name:	Date: ....

Topic	**Similarity and Ratios - Answers**
Notes	✓ Two figures are similar if they have the same shape. ✓ Two or more figures are similar if the corresponding angles are equal, and the corresponding sides are in proportion.
Example	*Following triangles are similar. What is the value of unknown side?* **Solution:** Find the corresponding sides and write a proportion: $\frac{4}{12} = \frac{x}{9}$. Now, use cross product to solve for x: $\frac{4}{12} = \frac{x}{9} \rightarrow 4 \times 9 = 12 \times x \rightarrow 36 = 12x$. Divide both sides by 12. Then: $5x = 40 \rightarrow \frac{36}{12} = \frac{12x}{12} \rightarrow x = 3$. The missing side is 3.

| **Your Turn!** | 1) 24

 2) 11

 3) 4

 4) 8

 5) 10

 6) 9 |

Name:	Date: ....

Topic	**Percent Problems**	

Notes	✓ In each percent problem, we are looking for the base, or part or the percent.
	 ✓ Use the following equations to find each missing section.
	      ○ Base = Part ÷ Percent
	      ○ Part = Percent × Base
	      ○ Percent = Part ÷ Base	

Examples	**18 *is what percent of* 30?** In this problem, we are looking for the percent. Use the following equation: $Percent = Part \div Base \rightarrow Percent = 18 \div 30 = 0.6 = 60\%$ **40 *is* 20% *of what number?*** Use the following formula: $Base = Part \div Percent \rightarrow Base = 40 \div 0.20 = 200$ 40 is 20% of 200.	
Your Turn!	1) What is 25 percent of 800?	2) 26 is what percent of 200?
	3) 60 is 5 percent of what number?	4) 48 is what percent of 300?
	5) 84 is 28 percent of what number?	6) 63 is what percent of 700?
	7) 96 is 24 percent of what number?	8) 40 is what percent of 800?

Name:	Date: ....

Topic	**Percent Problems – Answers**

Notes	✓ In each percent problem, we are looking for the base, or part or the percent.
	 ✓ Use the following equations to find each missing section.
	     ○ Base = Part ÷ Percent
	     ○ Part = Percent × Base
	     ○ Percent = Part ÷ Base

Examples	**18 *is what percent of* 30?** In this problem, we are looking for the percent. Use the following equation: $Percent = Part \div Base \rightarrow Percent = 18 \div 30 = 0.6 = 60\%$ **40 *is* 20% *of what number?*** Use the following formula: $Base = Part \div Percent \rightarrow Base = 40 \div 0.20 = 200$ 40 is 20% of 200.

Your Turn!	1) What is 25 percent of 800? 200	2) 26 is what percent of 200? 13%
	3) 60 is 5 percent of what number? 1,200	4) 48 is what percent of 300? 16%
	5) 84 is 28 percent of what number? 300	6) 63 is what percent of 700? 9%
	7) 96 is 24 percent of what number? 400	8) 40 is what percent of 800? 5%

| Name: .... | Date: .... |

Topic	**Percent of Increase and Decrease**

Notes	✓ Percent of change (increase or decrease) is a mathematical concept that represents the degree of change over time.
	
	✓ To find the percentage of increase or decrease:
	
	   1- New Number – Original Number
	
	   2- The result ÷ Original Number × 100
	
	Or use this formula: Percent of change = $\dfrac{new\ number - original\ number}{original\ number} \times 100$

Example	The price of a printer increases from \$40 to \$50. What is the percent increase? **Solution:** Percent of change = $\dfrac{new\ number - original\ number}{original\ number} \times 100 = \dfrac{50-40}{40} \times 100 = 25$ The percentage increase is 25. It means that the price of the printer increased 25%.
Your Turn!	1) In a class, the number of students has been increased from 32 to 36. What is the percentage increase? _____ % 2) The price of gasoline rose from \$4.50 to \$5.40 in one month. By what percent did the gas price rise? _____ % 3) A shirt was originally priced at \$65.00. It went on sale for \$52.00. What was the percent that the shirt was discounted? _____ % 4) Jason got a raise, and his hourly wage increased from \$40 to \$52. What is the percent increase? _____ %

Name:	**Date:** ....

Topic	**Percent of Increase and Decrease - Answers**
Notes	✓ Percent of change (increase or decrease) is a mathematical concept that represents the degree of change over time. ✓ To find the percentage of increase or decrease: 1- New Number − Original Number 2- The result ÷ Original Number × 100 Or use this formula: Percent of change = $\frac{new\ number\ -\ original\ number}{original\ number} \times 100$
Example	The price of a printer increases from \$40 to \$50. What is the percent increase? **Solution:** Percent of change = $\frac{new\ number\ -\ original\ number}{original\ number} \times 100 = \frac{50-40}{40} \times 100 = 25$ The percentage increase is 25. It means that the price of the printer increased 25%.
Your Turn!	1) In a class, the number of students has been increased from 32 to36. What is the percentage increase? 12.5% 2) The price of gasoline rose from \$4.50 to \$5.40 in one month. By what percent did the gas price rise? 20% 3) A shirt was originally priced at \$65.00. It went on sale for \$52.00. What was the percent that the shirt was discounted? 20% 4) Jason got a raise, and his hourly wage increased from \$40 to \$52. What is the percent increase? 30%

Name:	Date: ....

Topic	**Discount, Tax and Tip**
Notes	✓ Discount = Multiply the regular price by the rate of discount ✓ Selling price = original price − discount ✓ To find tax, multiply the tax rate to the taxable amount (income, property value, etc.) ✓ To find tip, multiply the rate to the selling price.
Example	The original price of a table is $300 and the tax rate is 6%. What is the final price of the table? **Solution:** First find the tax amount. To find tax: Multiply the tax rate to the taxable amount. Tax rate is 6% or 0.06. Then: $0.06 \times 300 = 18$. The tax amount is $18. Final price is: $300 + $18 = $318

Your Turn!	1) Original price of a chair: $300 Tax: 15%, Selling price: _____	2) Original price of a computer: $750 Discount: 20%, Selling price: _____
	3) Original price of a printer: $250 Tax: 10%, Selling price: _____	4) Original price of a sofa: $620 Discount: 25%, Selling price: _____
	5) Original price of a mattress: $800 Tax: 12%, Selling price: _____	6) Original price of a book: $150 Discount: 60%, Selling price: _____
	7) Restaurant bill: $35.00 Tip: 20%, Final amount: _____	8) Restaurant bill: $60.00 Tip: 25%, Final amount: _____

Name:	Date: ....

Topic	Discount, Tax and Tip - Answers	
Notes	✓ Discount = Multiply the regular price by the rate of discount ✓ Selling price = original price − discount ✓ To find tax, multiply the tax rate to the taxable amount (income, property value, etc.) ✓ To find tip, multiply the rate to the selling price.	
Example	***The original price of a table is $300 and the tax rate is 6%. What is the final price of the table?*** **Solution:** First find the tax amount. To find tax: Multiply the tax rate to the taxable amount. Tax rate is 6% or 0.06. Then: $0.06 \times 300 = 18$. The tax amount is $18. Final price is: $300 + $18 = $318	
Your Turn!	1) Original price of a chair: $300 Tax: 15%, Selling price: $345	2) Original price of a computer: $750 Discount: 20%, Selling price: $600
	3) Original price of a printer: $250 Tax: 10%, Selling price: $275	4) Original price of a sofa: $620 Discount: 25%, Selling price: $465
	5) Original price of a mattress: $800 Tax: 12%, Selling price: $896	6) Original price of a book: $150 Discount: 60%, Selling price: $60
	7) Restaurant bill: $35.00 Tip: 20%, Final amount: $42	8) Restaurant bill: $60.00 Tip: 25%, Final amount: $75

Name:	**Date:** ....

Topic	**Simple Interest**
Notes	✓ Simple Interest: The charge for borrowing money or the return for lending it. To solve a simple interest problem, use this formula: Interest = principal x rate x time ⇒ $I = p \times r \times t$
Example	***Find simple interest for*** $3,000$ ***investment at*** 5% ***for 4 years.*** **Solution:** Use Interest formula: $I = prt$ ($P = \$3,000$, r = 5% = 0.05 and $t = 4$) Then: $I = 3,000 \times 0.05 \times 4 = \$600$
Your Turn!	1) $250 at 4% for 3 years. Simple interest: $_____ 2) $3,300 at 5% for 6 years. Simple interest: $_____ 3) $720 at 2% for 5 years. Simple interest: $_____ 4) $2,200 at 8% for 4 years. Simple interest: $_____ 5) $1,800 at 3% for 2 years. Simple interest: $_____ 6) $530 at 4% for 5 years. Simple interest: $_____ 7) $7,000 at 5% for 3 months. Simple interest: $_____ 8) $880 at 5% for 9 months. Simple interest: $_____

| Name: .... | Date: .... |

Topic	**Simple Interest - Answers**
Notes	✓ Simple Interest: The charge for borrowing money or the return for lending it. To solve a simple interest problem, use this formula: Interest = principal x rate x time $\Rightarrow$ $I = p \times r \times t$
Example	**Find simple interest for** $3,000$ **investment at** 5% **for 4 years.** **Solution:** Use Interest formula: $I = prt$ ($P = \$3,000$, r $= 5\% = 0.05$ and $t = 4$) Then: $I = 3,000 \times 0.05 \times 4 = \$600$

Your Turn!	1) $250 at 4% for 3 years. Simple interest: $30	2) $3,300 at 5% for 6 years. Simple interest: $990
	3) $720 at 2% for 5 years. Simple interest: $72	4) $2,200 at 8% for 4 years. Simple interest: $704
	5) $1,800 at 3% for 2 years. Simple interest: $108	6) $530 at 4% for 5 years. Simple interest: $106
	7) $7,000 at 5% for 3 months. Simple interest: $87.50	8) $880 at 5% for 9 months. Simple interest: $33

Name:	Date: ....

Topic	**Simplifying Variable Expressions**
Notes	✓ In algebra, a variable is a letter used to stand for a number. The most common letters are: $x, y, z, a, b, c, m,$ and n. ✓ Algebraic expression is an expression contains integers, variables, and the math operations such as addition, subtraction, multiplication, division, etc. ✓ In an expression, we can combine "like" terms. (values with same variable and same power)
Example	**Simplify this expression.** $(6x + 8x + 9) =?$ Combine like terms. Then: $(6x + 8x + 4) = 14x + 9$ **(remember you cannot combine variables and numbers).**

Your Turn!	1) $5x + 2 - 2x =$	2) $4 + 7x + 3x =$
	3) $8x + 3 - 3x =$	4) $-2 - x^2 - 6x^2 =$
	5) $3 + 10x^2 + 2 =$	6) $8x^2 + 6x + 7x^2 =$
	7) $5x^2 - 12x^2 + 8x =$	8) $2x^2 - 2x - x + 5x^2 =$
	9) $4x - (12 - 30x) =$	10) $10x - (80x - 48) =$

| Name: .... | Date: .... |

Topic	**Simplifying Variable Expressions - Answers**
Notes	✓ In algebra, a variable is a letter used to stand for a number. The most common letters are: $x, y, z, a, b, c, m, and\ n$. ✓ Algebraic expression is an expression contains integers, variables, and the math operations such as addition, subtraction, multiplication, division, etc. ✓ In an expression, we can combine "like" terms. (values with same variable and same power)
Example	*Simplify this expression.* $(6x + 8x + 9) =?$ Combine like terms. Then: $(6x + 8x + 4) = 14x + 9$ *(remember you cannot combine variables and numbers).*

Your Turn!	1) $5x + 2 - 2x =$ $\qquad 3x + 2$	2) $4 + 7x + 3x =$ $\qquad 10x + 4$
	3) $8x + 3 - 3x =$ $\qquad 5x + 3$	4) $-2 - x^2 - 6x^2 =$ $\qquad -7x^2 - 2$
	5) $3 + 10x^2 + 2 =$ $\qquad 10x^2 + 5$	6) $8x^2 + 6x + 7x^2 =$ $\qquad 15x^2 + 6x$
	7) $5x^2 - 12x^2 + 8x =$ $\qquad -7x^2 + 8x$	8) $2x^2 - 2x - x + 5x^2 =$ $\qquad 72x^2 - 3x$
	9) $4x - (12 - 30x) =$ $\qquad 34x - 12$	10) $\quad 10x - (80x - 48) =$ $\qquad -70x - 48$

Name:	Date: ....

Topic	**Simplifying Polynomial Expressions**
Notes	✓ In mathematics, a polynomial is an expression consisting of variables and coefficients that involves only the operations of addition, subtraction, multiplication, and non–negative integer exponents of variables. $$P(x) = a_n x^n + a_{n-1} x^{n-1} + \dots + a_2 x^2 + a_1 x + a_0$$
Example	***Simplify this expression.*** $(2x^2 - x^4) - (4x^4 - x^2) =$ First use distributive property: → multiply $(-)$ into $(4x^4 - x^2)$ $(2x^2 - x^4) - (4x^4 - x^2) = 2x^2 - x^4 - 4x^4 + x^2$ Then combine "like" terms: $2x^2 - x^4 - 4x^4 + x^2 = 3x^2 - 5x^4$ And write in standard form: $3x^2 - 5x^4 = -5x^4 + 3x^2$

Your Turn!	1) $(2x^3 + 5x^2) - (12x + 2x^2) =$	2) $(2x^5 + 2x^3) - (7x^3 + 6x^2) =$
	3) $(12x^4 + 4x^2) - (2x^2 - 6x^4) =$	4) $14x - 3x^2 - 2(6x^2 + 6x^3) =$
	5) $(5x^3 - 3) + 5(2x^2 - 3x^3) =$	6) $(4x^3 - 2x) - 2(4x^3 - 2x^4) =$
	7) $2(4x - 3x^3) - 3(3x^3 + 4x^2) =$	8) $(2x^2 - 2x) - (2x^3 + 5x^2) =$

Name:	Date: ....

Topic	Simplifying Polynomial Expressions - Answers	
Notes	✓ In mathematics, a polynomial is an expression consisting of variables and coefficients that involves only the operations of addition, subtraction, multiplication, and non–negative integer exponents of variables. $$P(x) = a_n x^n + a_{n-1} x^{n-1} + \; ... \; + a_2 x^2 + a_1 x + a_0$$	
Example	**Simplify this expression.** $(2x^2 - x^4) - (4x^4 - x^2) =$ First use distributive property: → multiply (−) into $(4x^4 - x^2)$ $(2x^2 - x^4) - (4x^4 - x^2) = 2x^2 - x^4 - 4x^4 + x^2$ Then combine "like" terms: $2x^2 - x^4 - 4x^4 + x^2 = 3x^2 - 5x^4$ And write in standard form: $3x^2 - 5x^4 = -5x^4 + 3x^2$	
Your Turn!	1) $(2x^3 + 5x^2) - (12x + 2x^2) =$ $2x^3 + 3x^2 - 12x$ 3) $(12x^4 + 4x^2) - (2x^2 - 6x^4) =$ $18x^4 + 2x^2$ 5) $(5x^3 - 3) + 5(2x^2 - 3x^3) =$ $-10x^3 + 10x^2 - 3$ 7) $2(4x - 3x^3) - 3(3x^3 + 4x^2) =$ $-15x^3 - 12x^2 + 8x$	2) $(2x^5 + 2x^3) - (7x^3 + 6x^2) =$ $2x^5 - 5x^3 - 6x^2$ 4) $14x - 3x^2 - 2(6x^2 + 6x^3) =$ $-12x^3 - 15x^2 + 14x$ 6) $(4x^3 - 2x) - 2(4x^3 - 2x^4) =$ $4x^4 - 4x^3 - 2$ 8) $(2x^2 - 2x) - (2x^3 + 5x^2) =$ $-2x^3 - 3x^2 - 2x$

Name:	Date: ....

Topic	**Evaluating One Variable**
Notes	✓ To evaluate one variable expression, find the variable and substitute a number for that variable. ✓ Perform the arithmetic operations.
Example	**Find the value of this expression for** $x = -3$. $\quad -3x - 13$ **Solution:** Substitute -3 for x, then: $-3x - 13 = -3(-3) - 13 = 9 - 13 = -4$

Your Turn!	1) $x = -3 \Rightarrow 3x + 8 = $ _____	2) $x = 4 \Rightarrow 4(2x + 6) = $ _____
	3) $x = -1 \Rightarrow 6x + 4 = $ _____	4) $x = 7 \Rightarrow 6(5x + 3) = $ _____
	5) $x = 4 \Rightarrow 5(3x + 2) = $ ___	6) $x = 6 \Rightarrow 3(2x + 4) = $ _____
	7) $x = 3 \Rightarrow 7(3x + 1) = $ ___	8) $x = 8 \Rightarrow 3(3x + 7) = $ _____
	9) $x = 9 \Rightarrow 2(x + 9) = $ _____	10) $x = 7 \Rightarrow 2(4x + 5) = $ _____

Name:	Date: ....

Topic	Evaluating One Variable - Answers
Notes	✓ To evaluate one variable expression, find the variable and substitute a number for that variable. ✓ Perform the arithmetic operations.
Example	**Find the value of this expression for** $x = -3$. $-3x - 13$ **Solution:** Substitute -3 for x, then: $-3x - 13 = -3(-3) - 13 = 9 - 13 = -4$

Your Turn!	1) $x = -3 \Rightarrow 3x + 8 = -1$	2) $x = 4 \Rightarrow 4(2x + 6) = 56$
	3) $x = -1 \Rightarrow 6x + 4 = -2$	4) $x = 7 \Rightarrow 6(5x + 3) = 228$
	5) $x = 4 \Rightarrow 5(3x + 2) = 70$	6) $x = 6 \Rightarrow 3(2x + 4) = 48$
	7) $x = 3 \Rightarrow 7(3x + 1) = 70$	8) $x = 8 \Rightarrow 3(3x + 7) = 93$
	9) $x = 9 \Rightarrow 2(x + 9) = 36$	10) $x = 7 \Rightarrow 2(4x + 5) = 66$

Name:	Date: ....

Topic	**Evaluating Two Variables**
Notes	✓ To evaluate an algebraic expression, substitute a number for each variable. ✓ Perform the arithmetic operations to find the value of the expression.
Example	**Evaluate this expression for** $a = 4$ **and** $b = -2$. $5a - 6b$ **Solution:** Substitute 4 for a, and -2 for b, then: $5a - 6b = 5(4) - 6(-2) = 20 + 12 = 32$

Your Turn!

1) $-4a + 6b$, $a = 4$, $b = 3$

____

2) $5x + 3y$, $x = 2$, $y = -1$

____

3) $-5a + 3b$, $a = 2$, $b = -2$

____

4) $3x - 4y$, $x = 6$, $y = 2$

____

5) $2z + 14 + 6k$, $z = 5$, $k = 3$

____

6) $7a - (9 - 3b)$, $a = 1$, $b = 1$

____

7) $-6a + 3b$, $a = 4$, $b = 3$

____

8) $-2a + b$, $a = 6$, $b = 9$

____

9) $8x + 2y$, $x = 4$, $y = 5$

____

10) $z + 4 + 2k$, $z = 7$, $k = 4$

____

| Name: .... | Date: .... |

Topic	**Evaluating Two Variables - Answers**
Notes	✓ To evaluate an algebraic expression, substitute a number for each variable. ✓ Perform the arithmetic operations to find the value of the expression.
Example	*Evaluate this expression for* $a = 4$ *and* $b = -2$. $5a - 6b$ **Solution:** Substitute 4 for a, and -2 for b, then: $5a - 6b = 5(4) - 6(-2) = 20 + 12 = 32$

Your Turn!	1) $-4a + 6b$, $a = 4$, $b = 3$
	    2
		2) $5x + 3y$, $x = 2$, $y = -1$
		    7

	3) $-5a + 3b$, $a = 2$, $b = -2$
	    -16
		4) $3x - 4y$, $x = 6$, $y = 2$
		    10

	5) $2z + 14 + 6k$, $z = 5$,
	                $k = 3$
	   42
		6) $7a - (9 - 3b)$, $a = 1$,
		                $b = 1$
		   1

	7) $-6a + 3b$, $a = 4$, $b = 3$
	    -15
		8) $-2a + b$, $a = 6$, $b = 9$
		    -3

	9) $8x + 2y$, $x = 4$, $y = 5$
	    42
		10) $z + 4 + 2k$, $z = 7$, $k = 4$
		    19

Name:	Date: ....

Topic	**The Distributive Property**
Notes	✓ The distributive property (or the distributive property of multiplication over addition and subtraction) simplifies and solves expressions in the form of: $a(b + c)$ or $a(b - c)$ ✓ Distributive Property rule: $$a(b + c) = ab + ac$$
Example	***Simply.*** $(5)(2x - 8)$ **Solution:** Use Distributive Property rule: $a(b + c) = ab + ac$ $$(5)(2x - 8) = (5 \times 2x) + (5) \times (-8) = 10x - 40$$

Your Turn!		
	1) $(-2)(4 - 3x) =$	2) $(6 - 3x)(-7)$
	3) $6\,(5 - 9x) =$	4) $10(3 - 5x) =$
	5) $5(6 - 5x) =$	6) $(-2)(-5x + 3) =$
	7) $(8 - 9x)(5) =$	8) $(-16x + 15)(-3) =$
	9) $(-2x + 7)(3) =$	10) $(-18x + 25)(-2) =$

Name:	Date: ....

Topic	**The Distributive Property - Answers**	
Notes	✓ The distributive property (or the distributive property of multiplication over addition and subtraction) simplifies and solves expressions in the form of: $a(b + c)$ or $a(b - c)$ ✓ Distributive Property rule: $$a(b + c) = ab + ac$$	
Example	***Simply.*** $(5)(2x - 8)$ **Solution:** Use Distributive Property rule: $a(b + c) = ab + ac$ $$(5)(2x - 8) = (5 \times 2x) + (5) \times (-8) = 10x - 40$$	
Your Turn!	1) $(-2)(4 - 3x) = 6x - 8$	2) $(6 - 3x)(-7) = 21x - 42$
	3) $6(5 - 9x) = -54x + 30$	4) $10(3 - 5x) = -50x + 30$
	5) $5(6 - 5x) = -25x + 30$	6) $(-2)(-5x + 3) = 10x - 6$
	7) $(8 - 9x)(5) = -45x + 40$	8) $(-16x + 15)(-3) =$ $48x - 45$
	9) $(-2x + 7)(3) = -6x + 21$	10) $(-18x + 25)(-2) =$ $36x - 50$

Name:	Date: ....

Topic	One–Step Equations

Notes	✓ You only need to perform one Math operation in order to solve the one-step equations.
	 ✓ To solve one-step equation, find the inverse (opposite) operation is being performed.
	 ✓ The inverse operations are:
	     - Addition and subtraction
	     - Multiplication and division

Example	*Solve this equation.* $\quad x + 42 = 60 \Rightarrow x = ?$ Here, the operation is addition and its inverse operation is subtraction. To solve this equation, subtract 42 from both sides of the *equation:* $x + 42 - 42 = 60 - 42$ Then simplify: $x + 42 - 42 = 60 - 42 \Rightarrow x = 18$
Your Turn!	1) $x - 15 = 36 \Rightarrow x = $ ____ 2) $18 = 13 + x \Rightarrow x = $ ____ 3) $x - 22 = 54 \Rightarrow x = $ ____ 4) $x + 14 = 24 \Rightarrow x = $ ____ 5) $4x = 24 \Rightarrow x = $ ____ 6) $\frac{x}{6} = -3 \Rightarrow x = $ ____ 7) $99 = 11x \Rightarrow x = $ ____ 8) $\frac{x}{12} = 6 \Rightarrow x = $ ____

Name:	Date: ....

Topic	One–Step Equations - Answers

Notes	✓ You only need to perform one Math operation in order to solve the one-step equations.
	✓ To solve one-step equation, find the inverse (opposite) operation is being performed.
	✓ The inverse operations are:
	   - Addition and subtraction
	   - Multiplication and division

Example	**Solve this equation.** $x + 42 = 60 \Rightarrow x = ?$ Here, the operation is addition and its inverse operation is subtraction. To solve this equation, subtract 42 from both sides of the *equation:* $x + 42 - 42 = 60 - 42$ Then simplify: $x + 42 - 42 = 60 - 42 \Rightarrow x = 18$

Your Turn!	1) $x - 15 = 36 \Rightarrow x = 51$	2) $18 = 13 + x \Rightarrow x = 5$
	3) $x - 22 = 54 \Rightarrow x = 76$	4) $x + 14 = 24 \Rightarrow x = 10$
	5) $4x = 24 \Rightarrow x = 6$	6) $\frac{x}{6} = -3 \Rightarrow x = -18$
	7) $99 = 11x \Rightarrow x = 9$	8) $\frac{x}{12} = 6 \Rightarrow x = 72$

Name:	Date: ....

Topic	**Multi –Step Equations - Answers**
Notes	✓ Combine "like" terms on one side. ✓ Bring variables to one side by adding or subtracting. ✓ Simplify using the inverse of addition or subtraction. ✓ Simplify further by using the inverse of multiplication or division. ✓ Check your solution by plugging the value of the variable into the original equation.
Example	*Solve this equation for* x . $2x - 3 = 13$ **Solution:** The inverse of subtraction is addition. Add 3 to both sides of the equation. Then: $2x - 3 = 13 \Rightarrow 2x - 3 = 13 + 3$ $\Rightarrow 2x = 16$. Now, divide both sides by 2, then: $\frac{2x}{2} = \frac{16}{2} \Rightarrow x = 8$ Now, check the solution: $x = 8 \Rightarrow 2x - 3 = 13 \Rightarrow 2(8) - 3 = 13 \Rightarrow 16 - 3 = 13$ The answer $x = 8$ is correct.
Your Turn!	1) $4x - 12 = 8 \Rightarrow x =$ 2) $12 - 3x = -6 + 3x \Rightarrow x =$ 3) $3(4 - 2x) = 24 \Rightarrow x =$ 4) $15 + 5x = -7 - 6x \Rightarrow x =$ 5) $-2(5 + x) = 2 \Rightarrow x =$ 6) $12 - 2x = -3 - 5x \Rightarrow x =$ 7) $14 = -(x - 9) \Rightarrow x =$ 8) $11 - 4x = -4 - 3x \Rightarrow x =$

Name:	Date: ....

Topic	**Multi –Step Equations - Answers**	
Notes	✓ Combine "like" terms on one side. ✓ Bring variables to one side by adding or subtracting. ✓ Simplify using the inverse of addition or subtraction. ✓ Simplify further by using the inverse of multiplication or division. ✓ Check your solution by plugging the value of the variable into the original equation.	
Example	**Solve this equation for** x. $2x - 3 = 13$ **Solution:** The inverse of subtraction is addition. Add 3 to both sides of the equation. Then: $2x - 3 = 13 \Rightarrow 2x - 3 = 13 + 3$ $\Rightarrow 2x = 16$. Now, divide both sides by 2, then: $\frac{2x}{2} = \frac{16}{2} \Rightarrow x = 8$ Now, check the solution: $x = 8 \Rightarrow 2x - 3 = 13 \Rightarrow 2(8) - 3 = 13 \Rightarrow 16 - 3 = 13$ The answer $x = 8$ is correct.	
Your Turn!	1) $4x - 12 = 8 \Rightarrow x = 5$	2) $12 - 3x = -6 + 3x \Rightarrow x = 3$
	3) $3(4 - 2x) = 24 \Rightarrow x = -2$	4) $15 + 5x = -7 - 6x \Rightarrow x = -2$
	5) $-2(5 + x) = 2 \Rightarrow x = -6$	6) $12 - 2x = -3 - 5x \Rightarrow x = -5$
	7) $14 = -(x - 9) \Rightarrow x = -5$	8) $11 - 4x = -4 - 3x \Rightarrow x = 15$

Name:	**Date:** ....

Topic	**System of Equations**
Notes	✓ A system of equations contains two equations and two variables. For example, consider the system of equations: $x - 2y = -2, x + 2y = 10$ ✓ The easiest way to solve a system of equation is using the elimination method. The elimination method uses the addition property of equality. You can add the same value to each side of an equation. ✓ For the first equation above, you can add $x + 2y$ to the left side and 10 to the right side of the first equation: $x - 2y + (x + 2y) = -2 + 10$. Now, if you simplify, you get: $x - 2y + (x + 2y) = -2 + 10 \rightarrow 2x = 8 \rightarrow x = 4$. Now, substitute 4 for the x in the first equation: $4 - 2y = -2$. By solving this equation, $y = 3$
Example	What is the value of x and y in this system of equations? $\begin{cases} 3x - y = 7 \\ -x + 4y = 5 \end{cases}$ **Solution:** Solving System of Equations by Elimination: $\begin{array}{c} 3x - y = 7 \\ \underline{-x + 4y = 5} \end{array}$ Multiply the second equation by 3, then add it to the first equation. $\begin{array}{c} 3x - y = 7 \\ \underline{3(-x + 4y = 5)} \end{array} \Rightarrow \begin{array}{c} 3x - y = 7 \\ \underline{-3x + 12y = 15)} \end{array} \Rightarrow 11y = 22 \Rightarrow y = 2.$ Now, substitute 2 for y in the first equation and solve for x. $3x - (2) = 7 \Rightarrow 3x = 9 \Rightarrow x = 3$
Your Turn!	1) $-4x + 4y = 8$ $-4x + 2y = 6$ $x = \underline{\quad}$ $\qquad y = \underline{\quad}$ 2) $-5x + y = -3$ $3x - 8y = 24$ $x = \underline{\quad}$ $\qquad y = \underline{\quad}$ 3) $y = -2$ $4x - 3y = 8$ $x = \underline{\quad}$ $\qquad y = \underline{\quad}$ 4) $y = -3x + 5$ $5x - 4y = -3$ $x = \underline{\quad}$ $\qquad y = \underline{\quad}$ 5) $20x - 18y = -26$ $-10x + 6y = 22$ $x = \underline{\quad}$ $\qquad y = \underline{\quad}$ 6) $-9x - 12y = 15$ $2x - 6y = 14$ $x = \underline{\quad}$ $\qquad y = \underline{\quad}$

Name: ....

Date: ....

Topic	System of Equations- Answers
Notes	✓ A system of equations contains two equations and two variables. For example, consider the system of equations: $x - 2y = -2, x + 2y = 10$ ✓ The easiest way to solve a system of equation is using the elimination method. The elimination method uses the addition property of equality. You can add the same value to each side of an equation. ✓ For the first equation above, you can add $x + 2y$ to the left side and 10 to the right side of the first equation: $x - 2y + (x + 2y) = -2 + 10$. Now, if you simplify, you get: $x - 2y + (x + 2y) = -2 + 10 \rightarrow 2x = 8 \rightarrow x = 4$. Now, substitute 4 for the x in the first equation: $4 - 2y = -2$. By solving this equation, $y = 3$
Example	What is the value of x and y in this system of equations? $\begin{cases} 3x - y = 7 \\ -x + 4y = 5 \end{cases}$ **Solution:** Solving System of Equations by Elimination: $\begin{array}{l} 3x - y = 7 \\ \underline{-x + 4y = 5} \end{array}$ Multiply the second equation by 3, then add it to the first equation. $\begin{array}{l} 3x - y = 7 \\ \underline{3(-x + 4y = 5)} \end{array} \Rightarrow \begin{array}{l} 3x - y = 7 \\ \underline{-3x + 12y = 15)} \end{array} \Rightarrow 11y = 22 \Rightarrow y = 2$. Now, substitute 2 for y in the first equation and solve for x. $3x - (2) = 7 \Rightarrow 3x = 9 \Rightarrow x = 3$

Your Turn!		
1) $-4x + 4y = 8$ $-4x + 2y = 6$		2) $-5x + y = -3$ $3x - 8y = 24$
$x = -1$ $y = 1$		$x = 0$ $y = -3$
3) $y = -2$ $4x - 3y = 8$		4) $y = -3x + 5$ $5x - 4y = -3$
$x = \dfrac{1}{2}$ $y = -2$		$x = 1$ $y = 2$
5) $20x - 18y = -26$ $-10x + 6y = 22$		6) $-9x - 12y = 15$ $2x - 6y = 14$
$x = -4$ $y = -3$		$x = 1$ $y = -2$

Name:	Date: ....

Topic	**Graphing Single–Variable Inequalities**
Notes	✓ An inequality compares two expressions using an inequality sign. ✓ Inequality signs are: "less than" $<$, "greater than" $>$, "less than or equal to" $\leq$, and "greater than or equal to" $\geq$. ✓ To graph a single–variable inequality, find the value of the inequality on the number line. ✓ For less than ($<$) or greater than ($>$) draw open circle on the value of the variable. If there is an equal sign too, then use filled circle. ✓ Draw an arrow to the right for greater or to the left for less than.
Example	***Draw a graph for this inequality.*** $x < 5$ **Solution:** Since, the variable is less than 5, then we need to find 5 in the number lin and draw an open circle on it. Then, draw an arrow to the left. $\begin{array}{c}\longleftarrow\!\circ\!\!\longrightarrow \\ \scriptstyle -6\ -5\ -4\ -3\ -2\ -1\ \ 0\ \ 1\ \ 2\ \ 3\ \ 4\ \ 5\ \ 6 \end{array}$
Your Turn!	1) $x < 4$ ⟵─┼─┼─┼─┼─┼─┼─┼─┼─┼─┼─┼─┼─⟶ -6 -5 -4 -3 -2 -1 0 1 2 3 4 5 6 2) $x \geq -1$ ⟵─┼─┼─┼─┼─┼─┼─┼─┼─┼─┼─┼─┼─⟶ -6 -5 -4 -3 -2 -1 0 1 2 3 4 5 6 3) $x \geq -3$ ⟵─┼─┼─┼─┼─┼─┼─┼─┼─┼─┼─┼─┼─⟶ -6 -5 -4 -3 -2 -1 0 1 2 3 4 5 6 4) $x \leq 6$ ⟵─┼─┼─┼─┼─┼─┼─┼─┼─┼─┼─┼─┼─⟶ -6 -5 -4 -3 -2 -1 0 1 2 3 4 5 6 5) $x > -6$ ⟵─┼─┼─┼─┼─┼─┼─┼─┼─┼─┼─┼─┼─⟶ -6 -5 -4 -3 -2 -1 0 1 2 3 4 5 6 6) $2 > x$ ⟵─┼─┼─┼─┼─┼─┼─┼─┼─┼─┼─┼─┼─⟶ -6 -5 -4 -3 -2 -1 0 1 2 3 4 5 6 7) $-2 \leq x$ ⟵─┼─┼─┼─┼─┼─┼─┼─┼─┼─┼─┼─┼─⟶ -6 -5 -4 -3 -2 -1 0 1 2 3 4 5 6 8) $x > 0$ ⟵─┼─┼─┼─┼─┼─┼─┼─┼─┼─┼─┼─┼─⟶ -6 -5 -4 -3 -2 -1 0 1 2 3 4 5 6

Name:	Date: ....

Topic	Graphing Single–Variable Inequalities- Answers
Notes	✓ An inequality compares two expressions using an inequality sign. ✓ Inequality signs are: "less than" $<$, "greater than" $>$, "less than or equal to" $\leq$, and "greater than or equal to" $\geq$. ✓ To graph a single–variable inequality, find the value of the inequality on the number line. ✓ For less than ($<$) or greater than ($>$) draw open circle on the value of the variable. If there is an equal sign too, then use filled circle. ✓ Draw an arrow to the right for greater or to the left for less than.
Example	**Draw a graph for this inequality.** $x < 5$ **Solution:** Since, the variable is less than 5, then we need to find 5 in the number line and draw an open circle on it. Then, draw an arrow to the left. -6 -5 -4 -3 -2 -1 0 1 2 3 4 5 6
Your Turn!	1) $x < 4$ -6 -5 -4 -3 -2 -1 0 1 2 3 4 5 6 2) $x \geq -1$ -6 -5 -4 -3 -2 -1 0 1 2 3 4 5 6 3) $x \geq -3$ -6 -5 -4 -3 -2 -1 0 1 2 3 4 5 6 4) $x \leq 6$ -6 -5 -4 -3 -2 -1 0 1 2 3 4 5 6 5) $x > -6$ -6 -5 -4 -3 -2 -1 0 1 2 3 4 5 6 6) $2 > x$ -6 -5 -4 -3 -2 -1 0 1 2 3 4 5 6 7) $-2 \leq x$ -6 -5 -4 -3 -2 -1 0 1 2 3 4 5 6 8) $x > 0$ -6 -5 -4 -3 -2 -1 0 1 2 3 4 5 6

Name:	**Date:** ....

Topic	**One–Step Inequalities**
Notes	✓ Inequality signs are: "less than" $<$, "greater than" $>$, "less than or equal to" $\leq$, and "greater than or equal to" $\geq$. ✓ You only need to perform one Math operation in order to solve the one-step inequalities. ✓ To solve one-step inequalities, find the inverse (opposite) operation is being performed. ✓ For dividing or multiplying both sides by negative numbers, flip the direction of the inequality sign.
Example	***Solve this inequality.*** $x + 12 < 60 \Rightarrow$ _____ Here, the operation is addition and its inverse operation is subtraction. To solve this inequality, subtract 12 from both sides of the ***inequality:*** $x + 12 - 12 < 60 - 12$ Then simplify: $x < 48$
Your Turn!	1) $4x < -8 \Rightarrow$ _____ 2) $x + 6 > 28 \Rightarrow$ _____ 3) $-3x \geq 36 \Rightarrow$ _____ 4) $x - 16 \leq 4 \Rightarrow$ _____ 5) $\frac{x}{2} \geq -9 \Rightarrow$ _____ 6) $48 < 6x \Rightarrow$ _____ 7) $77 \leq 11x \Rightarrow$ _____ 8) $\frac{x}{4} > 9 \Rightarrow$ _____

Name:	Date: ....

Topic	One–Step Inequalities - Answers
Notes	✓ Inequality signs are: "less than" $<$, "greater than" $>$, "less than or equal to" $\leq$, and "greater than or equal to" $\geq$. ✓ You only need to perform one Math operation in order to solve the one-step inequalities. ✓ To solve one-step inequalities, find the inverse (opposite) operation is being performed. ✓ For dividing or multiplying both sides by negative numbers, flip the direction of the inequality sign.
Example	***Solve this inequality.*** $\quad x + 12 < 60 \Rightarrow$ _____ Here, the operation is addition and its inverse operation is subtraction. To solve this inequality, subtract 12 from both sides of the ***inequality:*** $x + 12 - 12 < 60 - 12$ Then simplify: $x < 48$

Your Turn!	1) $4x < -8 \Rightarrow x < -2$	2) $x + 6 > 28 \Rightarrow x > 22$
	3) $-3x \geq 36 \Rightarrow x \leq -12$	4) $x - 16 \leq 4 \Rightarrow x \leq 20$
	5) $\frac{x}{2} \geq -9 \Rightarrow x \geq -18$	6) $48 < 6x \Rightarrow 8 < x$
	7) $77 \leq 11x \Rightarrow 7 \leq x$	8) $\frac{x}{4} > 9 \Rightarrow x > 36$

| Name: .... | Date: .... |

Topic	**Multi –Step Inequalities**
Notes	✓ Isolate the variable. ✓ Simplify using the inverse of addition or subtraction. ✓ Simplify further by using the inverse of multiplication or division. ✓ For dividing or multiplying both sides by negative numbers, flip the direction of the inequality sign.
Example	*Solve this inequality.* $3x + 12 \leq 21$ **Solution:** First subtract 12 from both sides: $3x + 12 - 12 \leq 21 - 12$ Then simplify: $3x + 12 - 12 \leq 21 - 12 \rightarrow 3x \leq 9$ Now divide both sides by 3: $\frac{3x}{3} \leq \frac{9}{3} \rightarrow x \leq 3$
Your Turn!	1) $5x + 6 < 36 \rightarrow$ _____ 2) $2x - 8 \leq 6 \rightarrow$ _____ 3) $2x - 5 \leq 17 \rightarrow$ _____ 4) $14 - 7x \geq -7 \rightarrow$ _____ 5) $18 - 6x \geq -6 \rightarrow$ _____ 6) $2x - 18 \leq 16 \rightarrow$ _____ 7) $8 + 4x < 44 \rightarrow$ _____ 8) $5 - 4x < 17 \rightarrow$ _____

Name:	Date: ....

Topic	Multi –Step Inequalities - Answers
Notes	✓ Isolate the variable. ✓ Simplify using the inverse of addition or subtraction. ✓ Simplify further by using the inverse of multiplication or division. ✓ For dividing or multiplying both sides by negative numbers, flip the direction of the inequality sign.
Example	**Solve this inequality.** $3x + 12 \leq 21$ **Solution:** First subtract 12 from both sides: $3x + 12 - 12 \leq 21 - 12$ Then simplify: $3x + 12 - 12 \leq 21 - 12 \rightarrow 3x \leq 9$ Now divide both sides by 3: $\frac{3x}{3} \leq \frac{9}{3} \rightarrow x \leq 3$
Your Turn!	1) $5x + 6 < 36 \rightarrow x < 6$ 2) $2x - 8 \leq 6 \rightarrow x \leq 7$ 3) $2x - 5 \leq 17 \rightarrow x \leq 11$ 4) $14 - 7x \geq -7 \rightarrow x \leq 3$ 5) $18 - 6x \geq -6 \rightarrow x \leq 4$ 6) $2x - 18 \leq 16 \rightarrow x \leq 17$ 7) $8 + 4x < 44 \rightarrow x < 9$ 8) $5 - 4x < 17 \rightarrow x > -3$

Name:	**Date:** ....

Topic	**Finding Slope**
Notes	✓ The slope of a line represents the direction of a line on the coordinate plane. ✓ A line on coordinate plane can be drawn by connecting two points. ✓ To find the slope of a line, we need two points. ✓ The slope of a line with two points A (x_1, y_1) and B (x_2, y_2) can be found by using this formula: $\frac{y_2 - y_1}{x_2 - x_1} = \frac{rise}{run}$ ✓ The equation of a line is typically written as $y = mx + b$ where m is the slope and b is the y-intercept.
Examples	***Find the slope of the line through these two points:*** $(4, -12) \ and \ (9, 8)$. **Solution:** Slope $= \frac{y_2 - y_1}{x_2 - x_1}$. Let (x_1, y_1) be $(4, -12)$ and (x_2, y_2) be $(9, 8)$. ***Then:*** slope $= \frac{y_2 - y_1}{x_2 - x_1} = \frac{8 - (-12)}{9 - 4} = \frac{8 + 12}{5} = \frac{20}{5} = 4$ ***Find the slope of the line with equation*** $y = 5x - 6$ **Solution:** when the equation of a line is written in the form of $y = mx + b$, the slope is m. In this line: $y = 5x - 6$, the slope is 5.
Your Turn!	1) $(2, 3), (4, 7)$ Slope = ____
	2) $(-2, 2), (0, 4)$ Slope = ____
	3) $(4, -2), (2, 4)$ Slope = ____
	4) $(-4, -1), (0, 7)$ Slope = ____
	5) $y = 3x + 18$ Slope = ____
	6) $y = 12x - 3$ Slope = ____

Name:	Date: ....

Topic	**Finding Slope - Answers**
Notes	✓ The slope of a line represents the direction of a line on the coordinate plane. ✓ A line on coordinate plane can be drawn by connecting two points. ✓ To find the slope of a line, we need two points. ✓ The slope of a line with two points A (x_1, y_1) and B (x_2, y_2) can be found by using this formula: $\frac{y_2 - y_1}{x_2 - x_1} = \frac{rise}{run}$ ✓ The equation of a line is typically written as $y = mx + b$ where m is the slope and b is the y-intercept.
Examples	***Find the slope of the line through these two points:*** $(4, -12)$ *and* $(9, 8)$. **Solution:** Slope $= \frac{y_2 - y_1}{x_2 - x_1}$. Let (x_1, y_1) be $(4, -12)$ and (x_2, y_2) be $(9, 8)$. **Then:** slope $= \frac{y_2 - y_1}{x_2 - x_1} = \frac{8 - (-12)}{9 - 4} = \frac{8 + 12}{5} = \frac{20}{5} = 4$ ***Find the slope of the line with equation*** $y = 5x - 6$ **Solution:** when the equation of a line is written in the form of $y = mx + b$, the slope is m. In this line: $y = 5x - 6$, the slope is 5.
Your Turn!	1) $(2, 3), (4, 7)$ Slope $= 2$ 2) $(-2, 2), (0, 4)$ Slope $= 1$ 3) $(4, -2), (2, 4)$ Slope $= -3$ 4) $(-4, -1), (0, 7)$ Slope $= 2$ 5) $y = 3x + 18$ Slope $= 3$ 6) $y = 12x - 3$ Slope $= 12$

| Name: .... | Date: .... |

Topic	**Graphing Lines Using Slope–Intercept Form**
Notes	✓ Slope–intercept form of a line: given the slope m and the y–intercept (the intersection of the line and y-axis) b, then the equation of the line is: $$y = mx + b$$
Example	***Sketch the graph of*** $y = -2x - 1$. **Solution:** To graph this line, we need to find two points. When x is zero the value of y is -1. And when y is zero the value of x is $-\frac{1}{2}$. $$x = 0 \rightarrow y = -2(0) - 1 = -1, y = 0 \rightarrow 0$$ $$= -2x - 1 \rightarrow x = -\frac{1}{2}$$ Now, we have two points: $(0, -1)$ and $(-\frac{1}{2}, 0)$. Find the points and graph the line. Remember that the slope of the line is $-\frac{1}{2}$.
Your Turn!	1) $y = -4x + 1$ 2) $y = -x - 5$

| Name: .... | Date: .... |

Topic	**Graphing Lines Using Slope–Intercept Form - Answers**
Notes	✓ Slope–intercept form of a line: given the slope m and the y–intercept (the intersection of the line and y-axis) b, then the equation of the line is: $$y = mx + b$$
Example	**Sketch the graph of** $y = -2x - 1$. **Solution:** To graph this line, we need to find two points. When x is zero the value of y is -1. And when y is zero the value of x is $-\frac{1}{2}$. $$x = 0 \rightarrow y = -2(0) - 1 = -1, y = 0 \rightarrow 0$$ $$= -2x - 1 \rightarrow x = -\frac{1}{2}$$ Now, we have two points: $(0, -1)$ and $(-\frac{1}{2}, 0)$. Find the points and graph the line. Remember that the slope of the line is $-\frac{1}{2}$.
Your Turn!	1) $y = -4x + 1$ 2) $y = -x - 5$

Name:	Date: ....

Topic	**Writing Linear Equations**
Notes	✓ The equation of a line: $y = mx + b$ ✓ Identify the slope. ✓ Find the y–intercept. This can be done by substituting the slope and the coordinates of a point (x, y) on the line.
Example	**Write the equation of the line through** $(3, 1)$ **and** $(-1, 5)$. **Solution:** $Slop = \dfrac{y_2 - y_1}{x_2 - x_1} = \dfrac{5 - 1}{-1 - 3} = \dfrac{4}{-4} = -1 \rightarrow m = -1$ To find the value of b, you can use either points. The answer will be the same: $y = -x + b$ $(3, 1) \rightarrow 1 = -3 + b \rightarrow b = 4$ $(-1, 5) \rightarrow 5 = -(-1) + b \rightarrow b = 4$ The equation of the line is: $y = -x + 4$

Your Turn!	1) through: $(-2, 7), (1, 4)$ $y =$	2) through: $(6, 1), (5, 2)$ $y =$
	3) through: $(5, -1), (8, 2)$ $y =$	4) through: $(-2, 4), (4, -8)$ $y =$
	5) through: $(6, -5), (-5, 6)$ $y =$	6) through: $(4, -4), (-2, 8)$ $y =$
	7) through $(8, 8)$, Slope: 2 $y =$	8) through $(-7, 10)$, Slope: -2 $y =$

| Name: .... | Date: .... |

Topic	**Writing Linear Equations - Answers**
Notes	✓ The equation of a line: $y = mx + b$ ✓ Identify the slope. ✓ Find the y–intercept. This can be done by substituting the slope and the coordinates of a point (x, y) on the line.

Example	**Write the equation of the line through $(\mathbf{3}, \mathbf{1})$ and $(-\mathbf{1}, \mathbf{5})$.** **Solution:** $Slop = \frac{y_2 - y_1}{x_2 - x_1} = \frac{5 - 1}{-1 - 3} = \frac{4}{-4} = -1 \rightarrow m = -1$ To find the value of b, you can use either points. The answer will be the same: $y = -x + b$ $(3, 1) \rightarrow 1 = -3 + b \rightarrow b = 4$ $(-1, 5) \rightarrow 5 = -(-1) + b \rightarrow b = 4$ The equation of the line is: $y = -x + 4$

Your Turn!		
	1) through: $(-2, 7), (1, 4)$ $y = -x + 5$	2) through: $(6, 1), (5, 2)$ $y = -x + 7$
	3) through: $(5, -1), (8, 2)$ $y = x - 6$	4) through: $(-2, 4), (4, -8)$ $y = -2x$
	5) through: $(6, -5), (-5, 6)$ $y = -x + 1$	6) through: $(4, -4), (-2, 8)$ $y = -2x + 4$
	7) through $(8, 8)$, Slope: 2 $y = 2x - 8$	8) through $(-7, 10)$, Slope: -2 $y = -2x - 4$

Name:	Date: ....

Topic	**Finding Midpoint**
Notes	✓ The middle of a line segment is its midpoint. ✓ The Midpoint of two endpoints A (x_1, y_1) and B (x_2, y_2) can be found using this formula: $M(\frac{x_1+x_2}{2}, \frac{y_1+y_2}{2})$
Example	Find the midpoint of the line segment with the given endpoints. $(1, -2), (3, 6)$ **Solution:** Midpoint $= (\frac{x_1+x_2}{2}, \frac{y_1+y_2}{2}) \rightarrow (x_1, y_1) = (1, -2)$ and $(x_2, y_2) = (3, 6)$ Midpoint $= (\frac{1+3}{2}, \frac{-2+6}{2}) \rightarrow (\frac{4}{2}, \frac{4}{2}) \rightarrow M(2, 2)$

Your Turn!	1) $(6, 0), (-4, 2)$ ***Midpoint*** $= (\_\_, \_\_)$	2) $(4, -1), (2, 3)$ ***Midpoint*** $= (\_\_, \_\_)$
	3) $(-3, 4), (-5, 0)$ ***Midpoint*** $= (\_\_, \_\_)$	4) $(8, 1), (-4, 5)$ ***Midpoint*** $= (\_\_, \_\_)$
	5) $(6, 7), (-4, 5)$ ***Midpoint*** $= (\_\_, \_\_)$	6) $(2, -3), (2, 5)$ ***Midpoint*** $= (\_\_, \_\_)$
	7) $(7, 3), (-1, -7)$ ***Midpoint*** $= (\_\_, \_\_)$	8) $(3, 9), (-1, 5)$ ***Midpoint*** $= (\_\_, \_\_)$
	9) $(3, 4), (-7, -6)$ ***Midpoint*** $= (\_\_, \_\_)$	10) $(-5, 2), (11, -6)$ ***Midpoint*** $= (\_\_, \_\_)$

Name: ....

Date: ....

Topic	Finding Midpoint - Answers
Notes	✓ The middle of a line segment is its midpoint. ✓ The Midpoint of two endpoints A (x_1, y_1) and B (x_2, y_2) can be found using this formula: $M(\frac{x_1+x_2}{2}, \frac{y_1+y_2}{2})$
Example	Find the midpoint of the line segment with the given endpoints. $(\mathbf{1}, -\mathbf{2}), (\mathbf{3}, \mathbf{6})$ **Solution:** Midpoint $= (\frac{x_1+x_2}{2}, \frac{y_1+y_2}{2}) \rightarrow (x_1, y_1) = (1, -2)$ and $(x_2, y_2) = (3, 6)$ Midpoint $= (\frac{1+3}{2}, \frac{-2+6}{2}) \rightarrow (\frac{4}{2}, \frac{4}{2}) \rightarrow M(2, 2)$

Your Turn!	1) $(6, 0), (-4, 2)$ 2) $(4, -1), (2, 3)$
	
	    Midpoint $= (1, 1)$ *Midpoint* $= (3, 1)$
	
	3) $(-3, 4), (-5, 0)$ 4) $(8, 1), (-4, 5)$
	
	    Midpoint $= (-4, 2)$ *Midpoint* $= (2, 3)$
	
	5) $(6, 7), (-4, 5)$ 6) $(2, -3), (2, 5)$
	
	    Midpoint $= (1, 6)$ *Midpoint* $= (2, 1)$
	
	7) $(7, 3), (-1, -7)$ 8) $(3, 9), (-1, 5)$
	
	    Midpoint $= (3, -2)$ *Midpoint* $= (1, 7)$
	
	9) $(3, 4), (-7, -6)$ 10) $(-5, 2), (11, -6)$
	
	    Midpoint $= (-2, -1)$ *Midpoint* $= (3, -2)$

| Name: .... | Date: .... |

Topic	**Finding Distance of Two Points**
Notes	✓ Use this formula to find the distance of two points A (x_1, y_1) and B (x_2, y_2): $$d = \sqrt{(x_2 - x_1)^2 + (y_2 - y_1)^2}$$
Example	*Find the distance of two points* $(-1, 5)$ and $(4, -7)$. **Solution:** *Use distance of two points formula:* $d =$ $\sqrt{(x_2 - x_1)^2 + (y_2 - y_1)^2}$ $(x_1, y_1) = (-1, 5)$, and $(x_2, y_2) = (4, -7)$ Then: $d = \sqrt{(x_2 - x_1)^2 + (y_2 - y_1)^2} \to d =$ $\sqrt{(4 - (-1))^2 + (-7 - 5)^2} = \sqrt{(-5)^2 + (-12)^2} = \sqrt{25 + 144} =$ $\sqrt{169} = 13$
Your Turn!	1) $(6, 2), (-4, 2)$ *Distance* = ____ 2) $(2, -3), (2, 5)$ *Distance* = ____ 3) $(-5, 10), (7, 1)$ *Distance* = ____ 4) $(8, 1), (-4, 6)$ *Distance* = ____ 5) $(-3, 6), (-4, 5)$ *Distance* = ____ 6) $(4, -1), (14, 23)$ *Distance* = ____ 7) $(-3, 4), (-5, 0)$ *Distance* = ____ 8) $(3, 9), (-1, 5)$ *Distance* = ____

| Name: .... | Date: .... |

Topic	**Finding Distance of Two Points - Answers**
Notes	✓ Use this formula to find the distance of two points A (x_1, y_1) and B (x_2, y_2): $$d = \sqrt{(x_2 - x_1)^2 + (y_2 - y_1)^2}$$
Example	**Find the distance of two points** $(-1, 5)$ and $(4, -7)$. **Solution: Use distance of two points formula:** $d = \sqrt{(x_2 - x_1)^2 + (y_2 - y_1)^2}$ $(x_1, y_1) = (-1, 5)$, and $(x_2, y_2) = (4, -7)$ Then: $d = \sqrt{(x_2 - x_1)^2 + (y_2 - y_1)^2} \rightarrow d =$ $\sqrt{(4 - (-1))^2 + (-7 - 5)^2} = \sqrt{(-5)^2 + (-12)^2} = \sqrt{25 + 144} =$ $\sqrt{169} = 13$

Your Turn!	1) $(6, 2), (-4, 2)$ **Distance** $= 10$	2) $(2, -3), (2, 5)$ **Distance** $= 8$
	3) $(-5, 10), (7, 1)$ **Distance** $= 15$	4) $(8, 1), (-4, 6)$ **Distance** $= 13$
	5) $(-3, 6), (-4, 5)$ **Distance** $= \sqrt{2}$	6) $(4, -1), (14, 23)$ **Distance** $= 26$
	7) $(-3, 4), (-5, 0)$ **Distance** $= \sqrt{20} = 2\sqrt{5}$	8) $(3, 9), (-1, 5)$ **Distance** $= \sqrt{32} = 4\sqrt{2}$

Name:	**Date:** ....

Topic	**Multiplication Property of Exponents**
Notes	✓ Exponents are shorthand for repeated multiplication of the same number by itself. For example, instead of 2×2, we can write 2^2. For $3 \times 3 \times 3 \times 3$, we can write 3^4 ✓ In algebra, a variable is a letter used to stand for a number. The most common letters are: $x, y, z, a, b, c, m, and \ n$. ✓ Exponent's rules: $x^a \times x^b = x^{a+b}$, $\dfrac{x^a}{x^b} = x^{a-b}$ $\quad (x^a)^b = x^{a \times b} \qquad\qquad (xy)^a = x^a \times y^a \qquad \left(\dfrac{a}{b}\right)^c = \dfrac{a^c}{b^c}$
Example	***Multiply.*** $4x^3 \times 2x^2$ Use Exponent's rules: $x^a \times x^b = x^{a+b} \rightarrow x^3 \times x^2 = x^{3+2} = x^5$ Then: $4x^3 \times 2x^2 = 8x^5$

Your Turn!	1) $x^2 \times 3x =$	2) $5x^4 \times x^2 =$
	3) $3x^2 \times 4x^5 =$	4) $3x^2 \times 6xy =$
	5) $3x^5y \times 5x^2y^3 =$	6) $3x^2y^2 \times 5x^2y^8 =$
	7) $5x^2y \times 5x^2y^7 =$	8) $6x^6 \times 4x^9y^4 =$
	9) $8x^2y^5 \times 7x^5y^3 =$	10) $12x^6x^2 \times 3xy^5 =$

Name:	Date: ....

Topic	**Multiplication Property of Exponents - Answers**
Notes	✓ Exponents are shorthand for repeated multiplication of the same number by itself. For example, instead of 2×2, we can write 2^2. For $3 \times 3 \times 3 \times 3$, we can write 3^4 ✓ In algebra, a variable is a letter used to stand for a number. The most common letters are: $x, y, z, a, b, c, m,$ and n. ✓ Exponent's rules: $x^a \times x^b = x^{a+b}$, $\dfrac{x^a}{x^b} = x^{a-b}$ $(x^a)^b = x^{a \times b}$ $\qquad (xy)^a = x^a \times y^a$ $\qquad \left(\dfrac{a}{b}\right)^c = \dfrac{a^c}{b^c}$
Example	***Multiply.*** $4x^3 \times 2x^2$ Use Exponent's rules: $x^a \times x^b = x^{a+b} \rightarrow x^3 \times x^2 = x^{3+2} = x^5$ Then: $4x^3 \times 2x^2 = 8x^5$

Your Turn!	1) $x^2 \times 3x = 3x^3$	2) $5x^4 \times x^2 = 5x^6$
	3) $3x^2 \times 4x^5 = 12x^7$	4) $3x^2 \times 6xy = 18x^3y$
	5) $3x^5y \times 5x^2y^3 = 15x^7y^4$	6) $3x^2y^2 \times 5x^2y^8 = 15x^4y^{10}$
	7) $5x^2y \times 5x^2y^7 = 25x^4y^8$	8) $6x^6 \times 4x^9y^4 = 24x^{15}y^4$
	9) $8x^2y^5 \times 7x^5y^3 = 56x^7y^8$	10) $12x^6x^2 \times 3xy^5 = 36x^9y^5$

Name: ....

Date: ....

Topic	**Division Property of Exponents**
Notes	✓ For division of exponents use these formulas: $\frac{x^a}{x^b} = x^{a-b}$, $x \neq 0$ $\frac{x^a}{x^b} = \frac{1}{x^{b-a}}$, $x \neq 0$, $\qquad$ $\frac{1}{x^b} = x^{-b}$
Example	**Simplify.** $\frac{6x^3y}{36x^2y^3}$ First cancel the common factor: $6 \rightarrow \frac{6x^3y}{36x^2y^3} = \frac{x^3y}{6x^2y^3}$ Use Exponent's rules: $\frac{x^a}{x^b} = x^{a-b} \rightarrow \frac{x^3}{x^2} = x^{3-2} = x^1 = x$ Then: $\frac{6x^3y}{36x^2y^3} = \frac{xy}{9y^3} \rightarrow$ now cancel the common factor: $y \rightarrow \frac{xy}{6y^3} = \frac{x}{6y^2}$
Your Turn!	1) $\frac{3^7}{3^2} =$ $\qquad\qquad$ 2) $\frac{5x}{10x^3} =$ 3) $\frac{3x^3}{2x^5} =$ $\qquad\qquad$ 4) $\frac{12x^3}{14x^6} =$ 5) $\frac{12x^3}{9y^8} =$ $\qquad\qquad$ 6) $\frac{25xy^4}{5x^6y^2} =$ 7) $\frac{2x^4y^5}{7xy^2} =$ $\qquad\qquad$ 8) $\frac{16x^2y^8}{4x^3} =$ 9) $\frac{12x^4}{15x^7y^9} =$ $\qquad\qquad$ 10) $\frac{12yx^4}{10yx^8} =$

Name:	Date: ....

Topic	**Division Property of Exponents - Answers**	
Notes	✓ For division of exponents use following formulas: $\frac{x^a}{x^b} = x^{a-b}$, $x \neq 0$ $$\frac{x^a}{x^b} = \frac{1}{x^{b-a}}, x \neq 0, \qquad \frac{1}{x^b} = x^{-b}$$	
Example	*Simplify*. $\frac{6x^3y}{36x^2y^3}$ First cancel the common factor: $6 \rightarrow \frac{6x^3y}{36x^2y^3} = \frac{x^3y}{6x^2y^3}$ Use Exponent's rules: $\frac{x^a}{x^b} = x^{a-b} \rightarrow \frac{x^3}{x^2} = x^{3-2} = x^1 = x$ Then: $\frac{6x^3y}{36x^2y^3} = \frac{xy}{9y^3} \rightarrow$ now cancel the common factor: $y \rightarrow \frac{xy}{6y^3} = \frac{x}{6y^2}$	
Your Turn!	1) $\frac{3^7}{3^2} = 3^5$	2) $\frac{5x}{10x^3} = \frac{1}{2x^2}$
	3) $\frac{3x^3}{2x^5} = \frac{3}{2x^2}$	4) $\frac{12x^3}{14x^6} = \frac{6}{7x^3}$
	5) $\frac{12x^3}{9y^8} = \frac{4x^3}{3y^8}$	6) $\frac{25xy^4}{5x^6y^2} = \frac{5y^2}{x^5}$
	7) $\frac{2x^4y^5}{7xy^2} = \frac{2x^3y^3}{7}$	8) $\frac{16x^2y^8}{4x^3} = \frac{4y^8}{x}$
	9) $\frac{12x^4}{15x^7y^9} = \frac{4}{5x^3y^9}$	10) $\frac{12y^8x^4}{10y^2x^8} = \frac{6y^6}{5x^4}$

Name:	Date: ....

Topic	**Powers of Products and Quotients**
Notes	✓ For any nonzero numbers a and b and any integer x, $(ab)^x = a^x \times b^x$, $\left(\frac{a}{b}\right)^c = \frac{a^c}{b^c}$
Example	**Simplify.** $\left(\frac{2x^3}{x}\right)^2$ First cancel the common factor: $x \rightarrow \left(\frac{2x^3}{x}\right)^2 = \left(2x^2\right)^2$ Use Exponent's rules: $(ab)^x = a^x \times b^x$ Then: $\left(2x^2\right)^2 = (2)^2\left(x^2\right)^2 = 4x^4$

Your Turn!	1) $(4x^3 x^3)^2 =$	2) $(3x^3 \times 5x)^2 =$
	3) $(10x^{11}y^3)^2 =$	4) $(9x^7 y^5)^2 =$
	5) $(4x^4 y^6)^3 =$	6) $(3x \times 4y^3)^2 =$
	7) $\left(\frac{5x}{x^2}\right)^2 =$	8) $\left(\frac{x^4 y^4}{x^2 y^2}\right)^3 =$
	9) $\left(\frac{25x}{5x^6}\right)^2 =$	10) $\left(\frac{x^8}{x^6 y^2}\right)^2 =$

Name:	Date: ....

Topic	Powers of Products and Quotients - Answers	
Notes	✓ For any nonzero numbers a and b and any integer x, $$(ab)^x = a^x \times b^x, \left(\frac{a}{b}\right)^c = \frac{a^c}{b^c}$$	
Example	**Simplify.** $\left(\frac{2x^3}{x}\right)^2$ First cancel the common factor: $x \rightarrow \left(\frac{2x^3}{x}\right)^2 = (2x^2)^2$ Use Exponent's rules: $(ab)^x = a^x \times b^x$ Then: $(2x^2)^2 = (2)^2(x^2)^2 = 4x^4$	
Your Turn!	1) $(4x^3x^3)^2 = 16x^{12}$	2) $(3x^3 \times 5x)^2 = 225x^8$
	3) $(10x^{11}y^3)^2 = 100x^{22}y^6$	4) $(9x^7y^5)^2 = 81x^{14}y^{10}$
	5) $(4x^4y^6)^3 = 64\,x^{12}y^{18}$	6) $(3x \times 4y^3)^2 = 144x^2y^6$
	7) $\left(\frac{5x}{x^2}\right)^2 = \frac{25}{x^2}$	8) $\left(\frac{x^4y^4}{x^2y^2}\right)^3 = x^6y^6$
	9) $\left(\frac{25x}{5x^6}\right)^2 = \frac{25}{x^{10}}$	10) $\left(\frac{x^8}{x^6y^2}\right)^2 = \frac{x^4}{y^4}$

| Name: .... | Date: .... |

Topic	**Zero and Negative Exponents**
Notes	✓ A negative exponent is the reciprocal of that number with a positive exponent. $(3)^{-2} = \frac{1}{3^2}$ ✓ Zero-Exponent Rule: $a^0 = 1$, this means that anything raised to the zero power is 1. For example: $(28x^2 y)^0 = 1$
Example	***Evaluate.*** $\left(\frac{1}{3}\right)^{-2} =$ Use negative exponent's rule: $\left(\frac{1}{x^a}\right)^{-2} = (x^a)^2 \rightarrow \left(\frac{1}{3}\right)^{-2} = (3)^2 =$ Then: $(3)^2 = 9$
Your Turn!	1) $2^{-3} =$ 2) $3^{-3} =$ 3) $7^{-3} =$ 4) $1^{-3} =$ 5) $8^{-3} =$ 6) $4^{-4} =$ 7) $10^{-3} =$ 8) $7^{-4} =$ 9) $\left(\frac{1}{8}\right)^{-1} =$ 10) $\left(\frac{1}{5}\right)^{-2} =$

Name: ....   **Date:** ....

Topic	Zero and Negative Exponents - Answers
Notes	✓ A negative exponent is the reciprocal of that number with a positive exponent. $(3)^{-2} = \frac{1}{3^2}$ ✓ Zero-Exponent Rule: $a^0 = 1$, this means that anything raised to the zero power is 1. For example: $(28x^2y)^0 = 1$
Example	*Evaluate.* $\left(\frac{1}{3}\right)^{-2} =$ Use negative exponent's rule: $\left(\frac{1}{x^a}\right)^{-2} = (x^a)^2 \rightarrow \left(\frac{1}{3}\right)^{-2} = (3)^2 =$ Then: $(3)^2 = 9$

Your Turn!	1) $2^{-3} = \frac{1}{8}$	2) $3^{-3} = \frac{1}{27}$
	3) $7^{-3} = \frac{1}{343}$	4) $1^{-3} = 1$
	5) $8^{-3} = \frac{1}{512}$	6) $4^{-4} = \frac{1}{256}$
	7) $10^{-3} = \frac{1}{1,000}$	8) $7^{-4} = \frac{1}{2,401}$
	9) $\left(\frac{1}{8}\right)^{-1} = 8$	10) $\left(\frac{1}{5}\right)^{-2} = 25$

Name:	Date: ....

Topic	**Negative Exponents and Negative Bases**	
Notes	✓ Make the power positive. A negative exponent is the reciprocal of that number with a positive exponent. ✓ The parenthesis is important! 5^{-2} is not the same as $(-5)^{-2}$ $$(-5)^{-2} = -\frac{1}{5^2} \text{ and } (-5)^{-2} = +\frac{1}{5^2}$$	
Example	**Simplify.** $\left(-\frac{3x}{4yz}\right)^{-2} =$ Use negative exponent's rule: $\left(\frac{x^a}{x^b}\right)^{-2} = \left(\frac{x^b}{x^a}\right)^2 \rightarrow \left(-\frac{3x}{4yz}\right)^{-3} = \left(-\frac{4yz}{3x}\right)^3$ Now use exponent's rule: $\left(\frac{a}{b}\right)^c = \frac{a^c}{b^c} \rightarrow \left(-\frac{4yz}{3x}\right)^3 = \frac{4^3 y^3 z^3}{3^3 x^3} = \frac{64 y^3 z^3}{27 x^3}$	
Your Turn!	1) $-5x^{-2}y^{-3} =$	2) $20x^{-4}y^{-1} =$
	3) $14a^{-6}b^{-7} =$	4) $-12x^2 y^{-3} =$
	5) $-\frac{25}{x^{-6}} =$	6) $\frac{7b}{-9c^{-4}} =$
	7) $\frac{7ab}{a^{-3}b^{-1}} =$	8) $-\frac{5n^{-2}}{10p^{-3}} = -$
	9) $\frac{4ab^{-2}}{-3c^{-2}} =$	10) $\left(\frac{3a}{2c}\right)^{-2} =$

| Name: .... | Date: .... |

Topic	**Negative Exponents and Negative Bases - Answers**
Notes	✓ Make the power positive. A negative exponent is the reciprocal of that number with a positive exponent. ✓ The parenthesis is important! ✓ 5^{-2} is not the same as $(-5)^{-2}$ $$(-5)^{-2} = -\frac{1}{5^2} \text{ and } (-5)^{-2} = +\frac{1}{5^2}$$

| **Example** | **Simplify.** $\left(-\frac{3x}{4yz}\right)^{-2} =$

Use negative exponent's rule: $\left(\frac{x^a}{x^b}\right)^{-2} = \left(\frac{x^b}{x^a}\right)^2 \rightarrow \left(-\frac{3x}{4yz}\right)^{-3} = \left(-\frac{4yz}{3x}\right)^3$

Now use exponent's rule: $\left(\frac{a}{b}\right)^c = \frac{a^c}{b^c} \rightarrow \left(-\frac{4yz}{3x}\right)^3 = \frac{4^3y^3z^3}{3^3x^3} = \frac{64y^3z^3}{27x^3}$ |

Your Turn!		
	1) $-5x^{-2}y^{-3} = -\frac{5}{x^2 y^3}$	2) $20x^{-4}y^{-1} = \frac{20}{x^4 y}$
	3) $14a^{-6}b^{-7} = \frac{14}{a^6 b^7}$	4) $-12x^2y^{-3} = -\frac{12x^2}{y^3}$
	5) $-\frac{25}{x^{-6}} = -25x^6$	6) $\frac{7b}{-9c^{-4}} = -\frac{7bc^4}{9}$
	7) $\frac{7ab}{a^{-3}b^{-1}} = 7a^4b^2$	8) $-\frac{5n^{-2}}{10p^{-3}} = -\frac{p^3}{2n^2}$
	9) $\frac{4ab^{-2}}{-3c^{-2}} = -\frac{4ac^2}{3b^2}$	10) $\left(\frac{3a}{2c}\right)^{-2} = \frac{4c^2}{9a^2}$

Name:	Date: ....

Topic	**Scientific Notation**
Notes	✓ It is used to write very big or very small numbers in decimal form. ✓ In scientific notation all numbers are written in the form of: $$m \times 10^n$$ | **Decimal notation**	**Scientific notation** | |---|---| | 3 | 3×10^0 | | $-45,000$ | -4.5×10^4 | | 0.3 | 3×10^{-1} | | $2,122.456$ | 2.122456×10^3 |
Example	*Write 0.00054 in scientific notation.* First, move the decimal point to the right so that you have a number that is between 1 and 10. Then: $m = 5.4$ Now, determine how many places the decimal moved in step 1 by the power of 10. Then: $10^{-4} \rightarrow$ When the decimal moved to the right, the exponent is negative. Then: $0.00054 = 5.4 \times 10^{-4}$
Your Turn!	1) $0.000325 =$ 2) $0.000023 =$ 3) $52,000,000 =$ 4) $21,000 =$ 5) $3 \times 10^{-1} =$ 6) $5 \times 10^{-2} =$ 7) $1.2 \times 10^3 =$ 8) $2 \times 10^{-4} =$

Name:	Date: ....

Topic	**Scientific Notation - Answers**
Notes	✓ It is used to write very big or very small numbers in decimal form. ✓ In scientific notation all numbers are written in the form of: $$m \times 10^n$$ <table><tr><td>**Decimal notation**</td><td>**Scientific notation**</td></tr><tr><td>3</td><td>3×10^0</td></tr><tr><td>$-45,000$</td><td>-4.5×10^4</td></tr><tr><td>0.3</td><td>3×10^{-1}</td></tr><tr><td>2,122.456</td><td>2.122456×10^3</td></tr></table>
Example	**Write 0.00054 in scientific notation.** First, move the decimal point to the right so that you have a number that is between 1 and 10. Then: $m = 5.4$ Now, determine how many places the decimal moved in step 1 by the power of 10. Then: $10^{-4} \rightarrow$ When the decimal moved to the right, the exponent is negative. Then: $0.00054 = 5.4 \times 10^{-4}$

Your Turn!	1) $0.000325 = 3.25 \times 10^{-4}$	2) $0.00023 = 2.3 \times 10^{-5}$
	3) $52,000,000 = 5.2 \times 10^7$	4) $21,000 = 2.1 \times 10^4$
	5) $3 \times 10^{-1} = 0.3$	6) $5 \times 10^{-2} = 0.05$
	7) $1.2 \times 10^3 = 1,200$	8) $2 \times 10^{-4} = 0.0002$

Name:	Date: ....

Topic	Radicals
Notes	✓ If n is a positive integer and x is a real number, then: $\sqrt[n]{x} = x^{\frac{1}{n}}$, $\sqrt[n]{xy} = x^{\frac{1}{n}} \times y^{\frac{1}{n}}$, $\sqrt[n]{\frac{x}{y}} = \frac{x^{\frac{1}{n}}}{y^{\frac{1}{n}}}$, and $\sqrt[n]{x} \times \sqrt[n]{y} = \sqrt[n]{xy}$ ✓ A square root of x is a number r whose square is: $r^2 = x$ (r is a square root of x. ✓ To add and subtract radicals, we need to have the same values under the radical. For example: $\sqrt{3} + \sqrt{3} = 2\sqrt{3}$, $3\sqrt{5} - \sqrt{5} = 2\sqrt{5}$
Example	*Evaluate.* $\sqrt{32} + \sqrt{8} =$ **Solution:** Since we do not have the same values under the radical, we cannot add these two radicals. But we can simplify each radical. $\sqrt{32} = \sqrt{16} \times \sqrt{2} = 4\sqrt{2}$ and $\sqrt{8} = \sqrt{4} \times \sqrt{2} = 2\sqrt{2}$ Now, we have the same values under the radical. Then: $$\sqrt{32} + \sqrt{8} = 4\sqrt{2} + 2\sqrt{2} = 6\sqrt{2}$$

Your Turn!	1) $\sqrt{9} \times \sqrt{9} =$	2) $\sqrt{8} \times \sqrt{2} =$
	3) $\sqrt{3} \times \sqrt{27} =$	4) $\sqrt{32} \div \sqrt{2} =$
	5) $\sqrt{2} + \sqrt{8} =$	6) $\sqrt{27} - \sqrt{3} =$
	7) $4\sqrt{5} - 2\sqrt{5} =$	8) $3\sqrt{3} \times 2\sqrt{3} =$

Name: ....

Date: ....

Topic	Radicals - Answers

Notes	✓ If n is a positive integer and x is a real number, then: $\sqrt[n]{x} = x^{\frac{1}{n}}$, $\sqrt[n]{xy} = x^{\frac{1}{n}} \times y^{\frac{1}{n}}$, $\sqrt[n]{\frac{x}{y}} = \frac{x^{\frac{1}{n}}}{y^{\frac{1}{n}}}$, and $\sqrt[n]{x} \times \sqrt[n]{y} = \sqrt[n]{xy}$ ✓ A square root of x is a number r whose square is: $\boldsymbol{r^2 = x}$ ($\boldsymbol{r}$ is a square root of $\boldsymbol{x}$. ✓ To add and subtract radicals, we need to have the same values under the radical. For example: $\sqrt{3} + \sqrt{3} = 2\sqrt{3}$, $3\sqrt{5} - \sqrt{5} = 2\sqrt{5}$

Example	***Evaluate.*** $\sqrt{32} + \sqrt{8} =$ **Solution:** Since we do not have the same values under the radical, we cannot add these two radicals. But we can simplify each radical. $\sqrt{32} = \sqrt{16} \times \sqrt{2} = 4\sqrt{2}$ and $\sqrt{8} = \sqrt{4} \times \sqrt{2} = 2\sqrt{2}$ Now, we have the same values under the radical. Then: $$\sqrt{32} + \sqrt{8} = 4\sqrt{2} + 2\sqrt{2} = 6\sqrt{2}$$

Your Turn!	1) $\sqrt{9} \times \sqrt{9} = 9$	2) $\sqrt{8} \times \sqrt{2} = 4$
	3) $\sqrt{3} \times \sqrt{27} = 9$	4) $\sqrt{32} \div \sqrt{2} = 4$
	5) $\sqrt{2} + \sqrt{8} = 3\sqrt{2}$	6) $\sqrt{27} - \sqrt{3} = 2\sqrt{3}$
	7) $4\sqrt{5} - 2\sqrt{5} = 2\sqrt{5}$	8) $3\sqrt{3} \times 2\sqrt{3} = 18$

Name:	Date: ....

Topic	**Simplifying Polynomials**	
Notes	✓ Find "like" terms. (they have same variables with same power). ✓ Use "FOIL". (First–Out–In–Last) for binomials: $\qquad (x + a)(x + b) = x^2 + (b + a)x + ab$ ✓ Add or Subtract "like" terms using order of operation.	
Example	**Simplify this expression.** $(x + 3)(x - 8) =$ **Solution:** First apply FOIL method: $(a + b)(c + d) = ac + ad + bc + bd$ $(x + 3)(x - 8) = x^2 - 8x + 3x - 24$ Now combine like terms: $x^2 - 8x + 3x - 24 = x^2 - 5x - 24$	
Your Turn!	1) $-(2x - 4) =$ _____	2) $2(2x + 6) =$ _____
	3) $3x(3x - 4) =$ _____	4) $5x(2x + 8) =$ _____
	5) $-2x(5x + 6) + 5x =$ _____	6) $-4x(8x - 3) - x^2 =$ _____
	7) $(x + 4)(x + 5) =$ _____	8) $(x + 2)(x + 8) =$ _____
	9) $-4x^2 + 10x^3 + 5x^2 =$ _____	10) $-3x^5 + 10x^4 + 5x^5 =$ _____

Name:	**Date:** ....

Topic	**Simplifying Polynomials - Answers**
Notes	✓ Find "like" terms. (they have same variables with same power). ✓ Use "FOIL". (First–Out–In–Last) for binomials: $$(x + a)(x + b) = x^2 + (b + a)x + ab$$ ✓ Add or Subtract "like" terms using order of operation.
Example	***Simplify this expression.*** $(x + 3)(x - 8) =$ **Solution:** First apply FOIL method: $(a + b)(c + d) = ac + ad + bc + bd$ $(x + 3)(x - 8) = x^2 - 8x + 3x - 24$ Now combine like terms: $x^2 - 8x + 3x - 24 = x^2 - 5x - 24$

Your Turn!	1) $-(2x - 4) =$ $-2x + 4$	2) $2(2x + 6) =$ $4x + 12$
	3) $3x(3x - 4) =$ $9x^2 - 12x$	4) $5x(2x + 8) =$ $10x^2 + 40x$
	5) $-2x(5x + 6) + 5x =$ $-10x^2 - 7x$	6) $-4x(8x - 3) - x^2 =$ $-33x^2 + 12x$
	7) $(x + 4)(x + 5) =$ $x^2 + 9x + 20$	8) $(x + 2)(x + 8) =$ $x^2 + 10x + 16$
	9) $-4x^2 + 10x^3 + 5x^2 =$ $10x^3 + x^2$	10) $-3x^5 + 10x^4 + 5x^5 =$ $2x^5 + 10x^4$

Name:	Date: ....

Topic	**Adding and Subtracting Polynomials**
Notes	✓ Adding polynomials is just a matter of combining like terms, with some order of operations considerations thrown in. ✓ Be careful with the minus signs, and don't confuse addition and multiplication!
Example	***Simplify the expressions.*** $(3x^2 - 4x^3) - (5x^3 - 8x^2) =$ **Solution:** First use Distributive Property: $-(5x^3 - 8x^2) = -5x^3 + 8x^2$ $\rightarrow (3x^2 - 4x^3) - (5x^3 - 8x^2) = 3x^2 - 4x^3 - 5x^3 + 8x^2$ Now combine like terms: $3x^2 - 4x^3 - 5x^3 + 8x^2 = -9x^3 + 11x^2$

Your Turn!	1) $(x^2 - x) + (4x^2 - 5) =$ _____	2) $(2x^3 + x) - (x^3 + 2) =$ _____
	3) $(x^2 - 5x) + (6x^2 - 5) =$ _____	4) $(8x^2 - 2) - (3x^2 + 7) =$ _____
	5) $(3x^2 + 2) - (2 - 4x^2) =$ _____	6) $(x^3 + x^2) - (x^3 - 10) =$ _____
	7) $(3x^3 - 2x) - (x - x^3) =$ _____	8) $(x - 5x^4) - (2x^4 + 3x) =$ _____
	9) $(6x^3 + 5) - (4 - 5x^3) =$ _____	10) $(2x^2 + 5x^3) - (6x^3 + 7) =$ _____

Name: ....   **Date:** ....

Topic	Adding and Subtracting Polynomials - Answers
Notes	✓ Adding polynomials is just a matter of combining like terms, with some order of operations considerations thrown in. ✓ Be careful with the minus signs, and don't confuse addition and multiplication!
Example	**Simplify the expressions.** $(3x^2 - 4x^3) - (5x^3 - 8x^2) =$ **Solution:** First use Distributive Property: $-(5x^3 - 8x^2) = -5x^3 + 8x^2$ $\rightarrow (3x^2 - 4x^3) - (5x^3 - 8x^2) = 3x^2 - 4x^3 - 5x^3 + 8x^2$ Now combine like terms: $3x^2 - 4x^3 - 5x^3 + 8x^2 = -9x^3 + 11x^2$

Your Turn!		
	1) $(x^2 - x) + (4x^2 - 5) =$ $5x^2 - x - 5$	2) $(2x^3 + x) - (x^3 + 2) =$ $x^3 + x - 2$
	3) $(x^2 - 5x) + (6x^2 - 5) =$ $7x^2 - 5x - 5$	4) $(8x^2 - 2) - (3x^2 + 7) =$ $5x^2 - 9$
	5) $(3x^2 + 2) - (2 - 4x^2) =$ $7x^2$	6) $(x^3 + x^2) - (x^3 - 10) =$ $x^2 + 10$
	7) $(3x^3 - 2x) - (x - x^3) =$ $4x^3 - 3x$	8) $(x - 5x^4) - (2x^4 + 3x) =$ $7x^4 - 2x$
	9) $(6x^3 + 5) - (4 - 5x^3) =$ $11x^3 + 1$	10) $(2x^2 + 5x^3) - (6x^3 + 7) =$ $-x^3 + 2x^2 - 7$

Name:	Date: ....

Topic	**Multiplying Binomials**
Notes	✓A binomial is a polynomial that is the sum or the difference of two terms, each of which is a monomial. ✓To multiply two binomials, use "FOIL" method. (First–Out–In–Last) $(x + a)(x + b) = x \times x + x \times b + a \times x + a \times b = x^2 + bx + ax + ab$
Example	**Multiply.** $(x - 4)(x + 9) =$ **Solution:** Use "FOIL". (First–Out–In–Last): $(x - 4)(x + 9) = x^2 + 9x - 4x - 36$ Then simplify: $x^2 + 9x - 4x - 36 = x^2 + 5x - 36$

Your Turn!	1) $(x + 2)(x + 2) =$ _____	2) $(x + 3)(x + 2) =$ _____
	3) $(x - 3)(x + 4) =$ _____	4) $(x - 2)(x - 4) =$ _____
	5) $(x + 3)(x + 4) =$ _____	6) $(x + 5)(x + 4) =$ _____
	7) $(x - 6)(x - 5) =$ _____	8) $(x - 5)(x - 5) =$ _____
	9) $(x + 6)(x - 8) =$ _____	10) $(x - 9)(x + 7) =$ _____

| Name: .... | Date: .... |

Topic	**Multiplying Binomials - Answers**
Notes	✓A binomial is a polynomial that is the sum or the difference of two terms, each of which is a monomial. ✓To multiply two binomials, use "FOIL" method. (First–Out–In–Last) $(x + a)(x + b) = x \times x + x \times b + a \times x + a \times b = x^2 + bx + ax + ab$
Example	**Multiply.** $(x - 4)(x + 9) =$ **Solution:** Use "FOIL". (First–Out–In–Last): $(x - 4)(x + 9) = x^2 + 9x - 4x - 36$ Then simplify: $x^2 + 9x - 4x - 36 = x^2 + 5x - 36$

Your Turn!	1) $(x + 2)(x + 2) =$ $x^2 + 4x + 4$	2) $(x + 3)(x + 2) =$ $x^2 + 5x + 6$
	3) $(x - 3)(x + 4) =$ $x^2 + x - 12$	4) $(x - 2)(x - 4) =$ $x^2 - 6x + 8$
	5) $(x + 3)(x + 4) =$ $x^2 + 7x + 12$	6) $(x + 5)(x + 4) =$ $x^2 + 9x + 20$
	7) $(x - 6)(x - 5) =$ $x^2 - 11x + 30$	8) $(x - 5)(x - 5) =$ $x^2 - 10x + 25$
	9) $(x + 6)(x - 8) =$ $x^2 - 2x - 48$	10) $(x - 9)(x + 7) =$ $x^2 - 2x - 63$

Name:	Date: ....

Topic	**Multiplying and Dividing Monomials**
Notes	✓ When you divide or multiply two monomials you need to divide or multiply their coefficients and then divide or multiply their variables. ✓ In case of exponents with the same base, you need to subtract their powers. ✓ Exponent's rules: $$x^a \times x^b = x^{a+b}, \qquad \frac{x^a}{x^b} = x^{a-b}$$ $$\frac{1}{x^b} = x^{-b}, \quad (x^a)^b = x^{a \times b}$$ $$(xy)^a = x^a \times y^a$$
Example	***Divide expressions*** . $\dfrac{-18x^5y^6}{2xy^2} =$ **Solution:** Use exponents' division rule: $\dfrac{x^a}{x^b} = x^{a-b}, \dfrac{x^5}{x} = x^{5-1} = x^4$ and $\dfrac{y^6}{y^2} = y^4$ Then: $\dfrac{-18x^5y^6}{2xy^2} = -9x^4y^4$
Your Turn!	1) $(x^8y)(xy^2) =$ ____ 2) $(x^4y^3)(x^2y^3) =$ ____ 3) $(x^7y^4)(2x^5y^2) =$ ____ 4) $(3x^5y^4)(4x^6y^3) =$ ____ 5) $(-6x^8y^7)(4x^6y^9) =$ ____ 6) $(-2x^9y^3)(9x^7y^8) =$ ____ 7) $\dfrac{30x^8y^9}{6x^5y^4} =$ ____ 8) $\dfrac{-42x^{12}y^{16}}{7x^8y^9} =$ ____

Name: ....   **Date:** ....

Topic	**Multiplying and Dividing Monomials - Answers**	
Notes	✓ When you divide or multiply two monomials you need to divide or multiply their coefficients and then divide or multiply their variables. ✓ In case of exponents with the same base, you need to subtract their powers. ✓ Exponent's rules: $$x^a \times x^b = x^{a+b}, \qquad \frac{x^a}{x^b} = x^{a-b}$$ $$\frac{1}{x^b} = x^{-b}, \quad (x^a)^b = x^{a \times b}$$ $$(xy)^a = x^a \times y^a$$	
Example	***Divide expressions.*** $\frac{-18x^5y^6}{2xy^2} =$ **Solution:** Use exponents' division rule: $\frac{x^a}{x^b} = x^{a-b}$, $\frac{x^5}{x} = x^{5-1} = x^4$ and $\frac{y^6}{y^2} = y^4$ Then: $\frac{-18x^5y^6}{2xy^2} = -9x^4y^4$	

Your Turn!	1) $(x^8y)(xy^2) =$
	    x^9y^3
		2) $(x^4y^3)(x^2y^3) =$
		    x^6y^6

	3) $(x^7y^4)(2x^5y^2) =$
	    $2x^{12}y^6$
		4) $(3x^5y^4)(4x^6y^3) =$
		    $12x^{11}y^7$

	5) $(-6x^8y^7)(4x^6y^9) =$
	    $-24x^{14}y^{16}$
		6) $(-2x^9y^3)(9x^7y^8) =$
		    $-18x^{16}y^{11}$

	7) $\frac{30x^8y^9}{6x^5y^4} =$
	    $5x^3y^5$
		8) $\frac{-42x^{12}y^{16}}{7x^8y^9} =$
		    $-6x^4y^7$

Name:	Date: ....

Topic	**Multiplying a Polynomial and a Monomial**	
Notes	✓ When multiplying monomials, use the product rule for exponents. $x^a \times x^b = x^{a+b}$ ✓ When multiplying a monomial by a polynomial, use the distributive property. $$a \times (b + c) = a \times b + a \times c = ab + ac$$ $$a \times (b - c) = a \times b - a \times c = ab - ac$$	
Example	***Multiply expressions.*** $4x(5x - 8) =$ **Solution:** Use Distributive Property: $4x(5x - 8) = 4x \times 5x - 4x \times (8) =$ Now, simplify: $4x \times 5x - 4x \times (8) = 20x^2 - 32x$	
Your Turn!	1) $3x(2x + y) =$ _____	2) $x(x - 3y) =$ _____
	3) $-x(5x - 3y) =$ _____	4) $4x(x + 5y) =$ _____
	5) $-x(5x + 8y) =$ _____	6) $2x(6x - 7y) =$ _____
	7) $-3x(x^3 + 4y^2 - 6x) =$ _____	8) $7x(x^2 - 5y^2 + 4) =$ _____

Name:	Date: ....

Topic	Multiplying a Polynomial and a Monomial - Answers
Notes	✓ When multiplying monomials, use the product rule for exponents. $$x^a \times x^b = x^{a+b}$$ ✓ When multiplying a monomial by a polynomial, use the distributive property. $$a \times (b + c) = a \times b + a \times c = ab + ac$$ $$a \times (b - c) = a \times b - a \times c = ab - ac$$
Example	***Multiply expressions.*** $4x(5x - 8) =$ **Solution:** Use Distributive Property: $4x(5x - 8) = 4x \times 5x - 4x \times (8) =$ Now, simplify: $4x \times 5x - 4x \times (8) = 20x^2 - 32x$

Your Turn!	1) $3x(2x + y) =$ 2) $x(x - 3y) =$
	    $6x^2 + 3xy$ $x^2 - 3xy$
	
	3) $-x(5x - 3y) =$ 4) $4x(x + 5y) =$
	   $-5x^2 + 3xy$ $4x^2 + 20xy$
	
	5) $-x(5x + 8y) =$ 6) $2x(6x - 7y) =$
	   $-5x^2 - 8xy$ $12x^2 - 14xy$
	
	7) $-3x(x^3 + 4y^2 - 6x) =$ 8) $7x(x^2 - 5y^2 + 4) =$
	   $-3x^4 - 12xy^2 + 18x^2$ $7x^3 - 35xy^2 + 28x$

Name:	Date: ....

Topic	**Multiplying Monomials**
Notes	✓ A monomial is a polynomial with just one term: Examples: $5x$ or $7x^2yz^8$. ✓ When you multiply monomials, first multiply the coefficients (a number placed before and multiplying the variable) and then multiply the variables using multiplication property of exponents. $x^a \times x^b = x^{a+b}$
Example	*Multiply.* $(-3xy^4z^5) \times (2x^2y^5z^3) =$ **Solution:** Multiply coefficients and find same variables and use multiplication property of exponents: $x^a \times x^b = x^{a+b}$ $-3 \times 2 = -6$, $x \times x^2 = x^{1+2} = x^3$, $y^4 \times y^5 = y^{4+5} = y^9$, and $z^2 \times z^5 = z^{2+5} = z^7$ Then: $(-3xy^4z^5) \times (2x^2y^5z^3) = -6x^3y^9z^7$
Your Turn!	1) $2x^2 \times 4x^6 =$ _____ 2) $5x^7 \times 6x^4 =$ _____ 3) $-2x^2y^4 \times 6x^3y^2 =$ _____ 4) $-5x^5y \times 3x^3y^4 =$ _____ 5) $8x^7y^5 \times 5x^6y^3 =$ _____ 6) $-6x^7y^5 \times (-3x^9y^8) =$ _____ 7) $12x^8y^8z^4 \times 3x^4y^3z =$ _____ 8) $-8x^9y^7z^{11} \times 7x^6y^7z^5 =$ _____

Name:	Date: ....

Topic	**Multiplying Monomials**
Notes	✓ A monomial is a polynomial with just one term: Examples: $5x$ or $7x^2yz^8$. ✓ When you multiply monomials, first multiply the coefficients (a number placed before and multiplying the variable) and then multiply the variables using multiplication property of exponents. $x^a \times x^b = x^{a+b}$
Example	*Multiply.* $(-3xy^4z^5) \times (2x^2y^5z^3) =$ **Solution:** Multiply coefficients and find same variables and use multiplication property of exponents: $x^a \times x^b = x^{a+b}$ $-3 \times 2 = -6,\ x \times x^2 = x^{1+2} = x^3,\ y^4 \times y^5 = y^{4+5} = y^9,$ and $z^2 \times z^5 = z^{2+5} = z^7$ Then: $(-3xy^4z^5) \times (2x^2y^5z^3) = -6x^3y^9z^7$

Your Turn!	1) $2x^2 \times 4x^6 =$ $\quad 8x^8$	2) $5x^7 \times 6x^4 =$ $\quad 30x^{11}$
	3) $-2x^2y^4 \times 6x^3y^2 =$ $\quad -12x^5y^6$	4) $-5x^5y \times 3x^3y^4 =$ $\quad -15x^8y^5$
	5) $8x^7y^5 \times 5x^6y^3 =$ $\quad 40x^{13}y^8$	6) $-6x^7y^5 \times (-3x^9y^8) =$ $\quad 18x^{16}y^{13}$
	7) $12x^8y^8z^4 \times 3x^4y^3z =$ $36x^{12}y^{11}z^5$	8) $-8x^9y^7z^{11} \times 7x^6y^7z^5 =$ $-56x^{15}y^{14}z^{16}$

Name:	Date: ....

Topic	**Factoring Trinomials**
Notes	To factor trinomial, use of the following methods: ✓ "FOIL": $(x + a)(x + b) = x^2 + (b + a)x + ab$ ✓ "Difference of Squares": $$a^2 - b^2 = (a + b)(a - b)$$ $$a^2 + 2ab + b^2 = (a + b)(a + b)$$ $$a^2 - 2ab + b^2 = (a - b)(a - b)$$ ✓ "Reverse FOIL": $x^2 + (b + a)x + ab = (x + a)(x + b)$

Example	***Factor this trinomial.*** $x^2 + 12x + 32 =$
	    Solution: Break the expression into groups: $(x^2 + 4x) + (8x + 32)$
	    Now factor out x from $x^2 + 4x : x(x + 4)$, and factor out 8 from $8x + 32$:
	    $8(x + 4)$
	    Then: $(x^2 + 4x) + (8x + 32) = x(x + 4) + 8(x + 4)$
	    Now factor out like term: $(x + 4) \rightarrow (x + 4)(x + 8)$

Your Turn!	1) $x^2 + 6x + 9 =$ _____ 2) $x^2 + 5x + 6 =$ _____ 3) $x^2 + x + 12 =$ _____ 4) $x^2 - 6x + 8 =$ _____ 5) $x^2 + 7x + 12 =$ _____ 6) $x^2 + 12x + 32 =$ _____ 7) $x^2 - 11x + 30 =$ _____ 8) $x^2 - 14x + 45 =$ _____

Name: .... **Date:** ....

Topic	Factoring Trinomials - Answers
Notes	To factor trinomial, use of the following methods: ✓ "FOIL": $(x + a)(x + b) = x^2 + (b + a)x + ab$ ✓ "Difference of Squares": $$a^2 - b^2 = (a + b)(a - b)$$ $$a^2 + 2ab + b^2 = (a + b)(a + b)$$ $$a^2 - 2ab + b^2 = (a - b)(a - b)$$ ✓ "Reverse FOIL": $x^2 + (b + a)x + ab = (x + a)(x + b)$
Example	**Factor this trinomial.** $x^2 + 12x + 32 =$ **Solution:** Break the expression into groups: $(x^2 + 4x) + (8x + 32)$ Now factor out x from $x^2 + 4x : x(x + 4)$, and factor out 8 from $8x + 32$: $8(x + 4)$ Then: $(x^2 + 4x) + (8x + 32) = x(x + 4) + 8(x + 4)$ Now factor out like term: $(x + 4) \rightarrow (x + 4)(x + 8)$

Your Turn!		
	1) $x^2 + 6x + 9 =$ $(x + 3)(x + 3)$	2) $x^2 + 5x + 6 =$ $(x + 3)(x + 2)$
	3) $x^2 + x + 12 =$ $(x - 3)(x + 4)$	4) $x^2 - 6x + 8 =$ $(x - 2)(x - 4)$
	5) $x^2 + 7x + 12 =$ $(x + 3)(x + 4)$	6) $x^2 + 12x + 32 =$ $(x + 8)(x + 4)$
	7) $x^2 - 11x + 30 =$ $(x - 6)(x - 5)$	8) $x^2 - 14x + 45 =$ $(x - 9)(x - 5)$

Name: ....

Date: ....

Topic	**The Pythagorean Theorem**
Notes	✓ In any right triangle: $a^2 + b^2 = c^2$
Example	Right triangle ABC (not shown) has two legs of lengths 18 cm (AB) and 24 cm (AC). What is the length of the third side (BC)? **Solution:** Use Pythagorean Theorem: $a^2 + b^2 = c^2$ Then: $a^2 + b^2 = c^2 \rightarrow 18^2 + 24^2 = c^2 \rightarrow 324 + 576 = c^2$ $c^2 = 900 \rightarrow c = \sqrt{900} = 30 \; cm$
Your Turn!	1) _____ 2) _____ 3) _____ 4) _____

Name:	**Date:** ....

Topic	**The Pythagorean Theorem - Answers**
Notes	✓ In any right triangle: $a^2 + b^2 = c^2$
Example	Right triangle ABC (not shown) has two legs of lengths 18 cm (AB) and 24 cm (AC). What is the length of the third side (BC)? **Solution:** Use Pythagorean Theorem: $a^2 + b^2 = c^2$ Then: $a^2 + b^2 = c^2 \rightarrow 18^2 + 24^2 = c^2 \rightarrow 324 + 576 = c^2$ $c^2 = 900 \rightarrow c = \sqrt{900} = 30\ cm$
Your Turn!	1) 17 2) 30 3) 12 4) 9

Name:	Date: ....

Topic	**Triangles**

Notes	✓ In any triangle the sum of all angles is 180 degrees. ✓ Area of a triangle = $\frac{1}{2}(base \times height)$

Example

What is the area of the following triangle?

Solution:

Use the area formula: Area $= \frac{1}{2}(base \times height)$

$base = 16$ and $height = 6$

Area $= \frac{1}{2}(16 \times 6) = \frac{96}{2} = 48$

Your Turn!

5) _____

24

10

6) _____

18

28

7) _____

20

30

8) _____

32

46

| Name: .... | Date: .... |

Topic	**Triangles - Answers**
Notes	✓ In any triangle the sum of all angles is 180 degrees. ✓ Area of a triangle = $\frac{1}{2}(base \times height)$ h b
Example	**What is the area of the following triangle?** 6 16 **Solution:** Use the area formula: Area $= \frac{1}{2}(base \times height)$ $base = 16$ and $height = 6$ Area $= \frac{1}{2}(16 \times 6) = \frac{96}{2} = 48$
Your Turn!	5) 120 24, 10 6) 252 18, 28 7) 300 20, 30 8) 736 32, 46

Name: ....

Date: ....

Topic	Polygons

Notes

Perimeter of a square $= 4 \times side = 4s$

Perimeter of a rectangle $= 2(width + length)$

Perimeter of trapezoid $= a + b + c + d$

Perimeter of a regular hexagon $= 6a$

Perimeter of a parallelogram $= 2(l + w)$

Example

Find the perimeter of following regular hexagon.

Solution: Since the hexagon is regular, all sides are equal. Then: Perimeter of Hexagon $= 6 \times (one\ side)$
Perimeter of Hexagon $= 6 \times (one\ side) = 6 \times 9 = 54\ m$

Your Turn!

9) *(rectangle)* _____

9 in

15 in

10) _____

8 m

10 m   10 m

14 m

11) *(regular hexagon)* 5 m _____

12) *(parallelogram)* _____

10 in

16 in

Name: ....

Date: ....

Topic	Polygons - Answers
Notes	Perimeter of a square $= 4 \times side = 4s$ Perimeter of a rectangle $= 2(width + length)$ Perimeter of trapezoid $= a + b + c + d$ Perimeter of a regular hexagon $= 6a$ Perimeter of a parallelogram $= 2(l + w)$

Figures: square with side s; rectangle with width and length; trapezoid with sides a, b, c, d; regular hexagon with side a; parallelogram with sides l and w.

| **Example** | **Find the perimeter of following regular hexagon.**

Solution: Since the hexagon is regular, all sides are equal. Then: Perimeter of Hexagon $= 6 \times (one\ side)$
Perimeter of Hexagon $= 6 \times (one\ side) = 6 \times 9 = 54\ m$

Hexagon with sides labeled 9 m, 9 m, 9 m |

| **Your Turn!** | 9) *(rectangle)* 48 in

Rectangle: 15 in by 9 in | 10)　42 m

Trapezoid: 8 m (top), 10 m, 10 m (sides), 14 m (bottom) |
| | 11)　*(regular hexagon)* 30 m

Hexagon with side 5 m | 12)　*(parallelogram)* 52 in

Parallelogram: 16 in, 10 in |

Name: ....   **Date:** ....

Topic	Circles

Notes

✓ In a circle, variable r is usually used for the radius and d for diameter and π is about 3.14.
✓ *Area of a circle* $= \pi r^2$
✓ *Circumference of a circle* $= 2\pi r$

r

Example

Find the area of the circle.

    Solution:
    Use area formula: $Area = \pi r^2$
    $r = 2\ in \rightarrow Area = \pi(2)^2 = 4\pi,\ \pi = 3.14$
    Then: $Area = 4 \times 3.14 = 12.56\ in^2$

$2\ in$

Your Turn!

Find the area of each circle. $(\pi = 3.14)$

1) _____

6 cm

2) _____

10 in

Find the Circumference of each circle. $(\pi = 3.14)$

3) _____

8 cm

4) _____

6 m

Name: ....   **Date:** ....

Topic	Circles - Answers
Notes	✓ In a circle, variable r is usually used for the radius and d for diameter and π is about 3.14. ✓ $Area\ of\ a\ circle = \pi r^2$ ✓ $Circumference\ of\ a\ circle = 2\pi r$ r
Example	**Find the area of the circle.** Solution: Use area formula: $Area = \pi r^2$ $r = 2\ in \rightarrow Area = \pi(2)^2 = 4\pi,\ \pi = 3.14$ **Then:** $Area = 4 \times 3.14 = 12.56\ in^2$ $2\ in$
Your Turn!	**Find the area of each circle.** ($\pi = 3.14$) 1) $113.04\ cm^2$ 2) $314\ in^2$ 6 cm 10 in **Find the Circumference of each circle.** ($\pi = 3.14$) 3) $50.24\ cm$ 4) $37.68\ m$ 8 cm 6 m

Name:	Date: ....

Topic	**Cubes**
Notes	✓ A cube is a three-dimensional solid object bounded by six square sides. ✓ Volume is the measure of the amount of space inside of a solid figure, like a cube, ball, cylinder or pyramid. ✓ Volume of a cube = $(one\ side)^3$ ✓ surface area of cube = $6 \times (one\ side)^2$
Example	**Find the volume and surface area of the following cube.** 15 cm **Solution:** Use volume formula: $volume = (one\ side)^3$ Then: $volume = (one\ side)^3 = (15)^3 = 3{,}375\ cm^3$ Use surface area formula: $surface\ area\ of\ cube: 6(one\ side)^2 = 6(15)^2 = 6(225) = 1{,}350\ cm^2$
Your Turn!	**Find the volume of each cube.** 1) _____ 11 in 2) _____ 13 ft 3) _____ 14 cm 4) _____ 30 m

Name: ....

Date: ....

Topic	Cubes - Answers
Notes	✓ A cube is a three-dimensional solid object bounded by six square sides. ✓ Volume is the measure of the amount of space inside of a solid figure, like a cube, ball, cylinder or pyramid. ✓ Volume of a cube $= (one\ side)^3$ ✓ surface area of cube $= 6 \times (one\ side)^2$

Example	**Find the volume and surface area of the following cube.** $15\ cm$ **Solution:** Use volume formula: $volume = (one\ side)^3$ Then: $volume = (one\ side)^3 = (15)^3 = 3,375\ cm^3$ Use surface area formula: $surface\ area\ of\ cube: 6(one\ side)^2 = 6(15)^2 = 6(225) = 1,350\ cm^2$

Your Turn!	**Find the volume of each cube.** 1) $1,331\ in^3$ $11\ in$ 2) $2,197\ ft^3$ $13\ ft$ 3) $2,744\ cm^3$ $14\ cm$ 4) $27,000\ m^3$ $30\ m$

| Name: .... | Date: .... |

Topic	**Trapezoids**
Notes	✓ A quadrilateral with at least one pair of parallel sides is a trapezoid. ✓ Area of a trapezoid $= \frac{1}{2}h(b_1 + b_2)$ 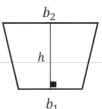
Example	***Calculate the area of the trapezoid.*** **Solution:** Use area formula: $A = \frac{1}{2}h(b_1 + b_2)$ $b_1 = 8\ cm$, $b_2 = 12\ cm$ and $h = 14\ cm$ Then: $A = \frac{1}{2}(14)(12 + 8) = 7(20) = 140\ cm^2$ 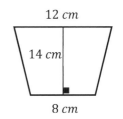
Your Turn!	1) _____ 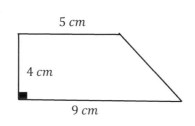 2) _____ 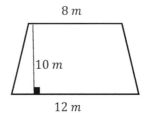 3) _____ 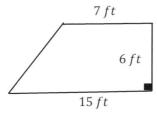 4) _____ 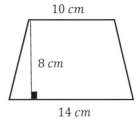

Name: .... **Date:** ....

Topic	Trapezoids - Answers

Notes
- ✓ A quadrilateral with at least one pair of parallel sides is a trapezoid.
- ✓ Area of a trapezoid $= \frac{1}{2}h(b_1 + b_2)$

b_2, h, b_1

Example

Calculate the area of the trapezoid.
Solution:
Use area formula: $A = \frac{1}{2}h(b_1 + b_2)$
$b_1 = 8\ cm$, $b_2 = 12\ cm$ and $h = 14\ cm$
Then: $A = \frac{1}{2}(14)(12 + 8) = 7(20) = 140\ cm^2$

12 cm, 14 cm, 8 cm

Your Turn!

1) $28\ cm^2$
5 cm, 4 cm, 9 cm

2) $100\ m^2$
8 m, 10 m, 12 m

3) $66\ ft^2$
7 ft, 6 ft, 15 ft

4) $96\ cm^2$
10 cm, 8 cm, 14 cm

Name:	Date: ....

Topic	**Rectangular Prisms**	
Notes	✓ A solid 3-dimensional object which has six rectangular faces. ✓ Volume of a Rectangular prism = **Length × Width × Height** $Volume = l \times w \times h$ $Surface\ area = 2(wh + lw + lh)$	
Example	**Find the volume and surface area of rectangular prism.** **Solution:** Use volume formula: $Volume = l \times w \times h$ Then: $Volume = 4 \times 2 \times 6 = 48\ m^3$ Use surface area formula: $Surface\ area = 2(wh + lw + lh)$ Then: $Surface\ area = 2\big((2 \times 6) + (4 \times 2) + (4 \times 6)\big)$ $= 2(12 + 8 + 24) = 2(44) = 88\ m^2$	
Your Turn!	**Find the surface area of each Rectangular Prism.** 1) _____ 6 ft 10 ft 4 ft 3) _____ 12 m 18 m 10 m	2) _____ 8 cm 16 cm 6 cm 4) _____ 20 in 15 in 12 in

Name:	Date: ....

Topic	**Rectangular Prisms - Answers**	
Notes	✓ A solid 3-dimensional object which has six rectangular faces. ✓ Volume of a Rectangular prism = **Length × Width × Height** $Volume = l \times w \times h$ $Surface\ area = 2(wh + lw + lh)$	

Example

Find the volume and surface area of rectangular prism.

Solution:

Use volume formula: $Volume = l \times w \times h$

Then: $Volume = 4 \times 2 \times 6 = 48\ m^3$

Use surface area formula: $Surface\ area = 2(wh + lw + lh)$

Then: $Surface\ area = 2\big((2 \times 6) + (4 \times 2) + (4 \times 6)\big)$

$$= 2(12 + 8 + 24) = 2(44) = 88\ m^2$$

Your Turn!

Find the surface area of each Rectangular Prism.

1) $248\ ft^2$

    6 ft   10 ft   4 ft

2) $544\ cm^2$

    8 cm   16 cm   6 cm

3) $1,032\ m^2$

    12 m   18 m   10 m

4) $1,440\ in^2$

    20 in   15 in   12 in

| Name: .... | Date: .... |

| Topic | **Cylinder** |

| **Notes** | ✓ A cylinder is a solid geometric figure with straight parallel sides and a circular or oval cross section.
 ✓ *Volume of Cylinder Formula* $= \pi(radius)^2 \times height$ $\pi = 3.14$
 ✓ *Surface area of a cylinder* $= 2\pi r^2 + 2\pi rh$

 height <br> radius |

| **Example** | ***Find the volume and Surface area of the follow Cylinder.***

 Solution:
 Use volume formula: $Volume = \pi(radius)^2 \times height$
 Then: $Volume = \pi(3)^2 \times 12 = 9\pi \times 12 = 108\pi$
 $\pi = 3.14$ **then:** $Volume = 108\pi = 339.12 \; cm^3$
 Use surface area formula: $Surface\; area = 2\pi r^2 + 2\pi rh$
 Then: $2\pi(3)^2 + 2\pi(3)(12) = 2\pi(9) + 2\pi(36) = 18\pi + 72\pi = 90\pi$
 $\pi = 3.14$ **Then:** $Surface\; area = 90 \times 3.14 = 282.6 \; cm^2$

 12 cm
 3 cm |

| **Your Turn!** | ***Find the volume of each Cylinder.*** $(\pi = 3.14)$

 1) _____   2) _____
 10 in   2 in   14 m   5 m

 Find the Surface area of each Cylinder. $(\pi = 3.14)$

 3) _____   4) _____
 15 ft   9 ft   20 cm   12 cm |

| Name: .... | Date: .... |

Topic	Cylinder - Answers
Notes	✓ A cylinder is a solid geometric figure with straight parallel sides and a circular or oval cross section. ✓ *Volume of Cylinder Formula* $= \pi(radius)^2 \times height \; \pi = 3.14$ ✓ *Surface area of a cylinder* $= 2\pi r^2 + 2\pi rh$

Example

Find the volume and Surface area of the follow Cylinder.

Solution:
Use volume formula: $Volume = \pi(radius)^2 \times height$
Then: $Volume = \pi(3)^2 \times 12 = 9\pi \times 12 = 108\pi$
$\pi = 3.14$ **then:** $Volume = 108\pi = 339.12 \; cm^3$
Use surface area formula: $Surface \; area = 2\pi r^2 + 2\pi rh$
Then: $2\pi(3)^2 + 2\pi(3)(12) = 2\pi(9) + 2\pi(36) = 18\pi + 72\pi = 90\pi$
$\pi = 3.14$ **Then:** $Surface \; area = 90 \times 3.14 = 282.6 \; cm^2$

(12 cm, 3 cm)

Your Turn!

Find the volume of each Cylinder. ($\pi = 3.14$)

1) $125.6 \; in^3$

(10 in, 2 in)

2) $1,099 \; m^3$

(14 m, 5 m)

Find the Surface area of each Cylinder. ($\pi = 3.14$)

3) $1,356.48 \; ft^2$

(15 ft, 9 ft)

4) $2,411.52 \; cm^2$

(20 cm, 12 cm)

Name:	Date: ....

Topic	**Mean, Median, Mode, and Range of the Given Data**
Notes	✓ Mean: $\dfrac{sum\ of\ the\ data}{total\ number\ of\ data\ entires}$ ✓ Mode: value in the list that appears most often. ✓ Median: is the middle number of a group of numbers that have been arranged in order by size. ✓ Range: the difference of largest value and smallest value in the list.
Example	**Find the mode and median of these numbers?** $16, 10, 6, 3, 1, 16, 2, 4$ **Solution:** Mode: value in the list that appears most often. Number 16 is the value in the list that appears most often (there are two number 16). To find median, write the numbers in order: $1, 2, 3, 4, 6, 10, 16, 16$ Number 4 and 6 are in the middle. Find their average: $\dfrac{4+6}{2} = \dfrac{10}{2} = 5$ The median is 5.
Your Turn!	**1)** $3, 2, 4, 8, 3, 10$ Mode: _____ Range: _____ Mean: _____ Median: _____ **3)** $5, 4, 3, 2, 9, 5, 6, 8, 12$ Mode: _____ Range: _____ Mean: _____ Median: _____	**2)** $6, 3, 2, 9, 5, 7, 2, 14$ Mode: _____ Range: _____ Mean: _____ Median: _____ **4)** $12, 6, 8, 6, 9, 6, 4, 13$ Mode: _____ Range: _____ Mean: _____ Median: _____

Name:	**Date:** ....

Topic	**Mean, Median, Mode, and Range of the Given Data - Answers**
Notes	✓ Mean: $\dfrac{sum\ of\ the\ data}{total\ number\ of\ data\ entires}$ ✓ Mode: value in the list that appears most often. ✓ Median: is the middle number of a group of numbers that have been arranged in order by size. ✓ Range: the difference of largest value and smallest value in the list.
Example	**Find the mode and median of these numbers?** $16, 10, 6, 3, 1, 16, 2, 4$ **Solution:** Mode: value in the list that appears most often. Number 16 is the value in the list that appears most often (there are two number 16). To find median, write the numbers in order: $1, 2, 3, 4, 6, 10, 16, 16$ Number 4 and 6 are in the middle. Find their average: $\dfrac{4+6}{2} = \dfrac{10}{2} = 5$ The median is 5.

Your Turn!

1) $3, 2, 4, 8, 3, 10$

Mode: 3 Range: 8

Mean: 5 Median: 3.5

2) $6, 3, 2, 9, 5, 7, 2, 14$

Mode: 2 Range: 12

Mean: 6 Median: 5.5

3) $5, 4, 3, 2, 9, 5, 6, 8, 12$

Mode: 5 Range: 10

Mean: 6 Median: 5

4) $12, 6, 8, 6, 9, 6, 4, 13$

Mode: 6 Range: 9

Mean: 8 Median: 7

Name:	Date: ....

Topic	**Probability Problems**
Notes	✓ Probability is the likelihood of something happening in the future. It is expressed as a number between zero (can never happen) to 1 (will always happen). ✓ Probability can be expressed as a fraction, a decimal, or a percent. ✓ Probability formula: $Probability = \frac{number\ of\ desired\ outcomes}{number\ of\ total\ outcomes}$
Example	*If there are 3 green balls, 4 red balls, and 10 blue balls in a basket, what is the probability that Jason will pick out a red ball from the basket?* **Solution:** There are 4 red ball and 17 are total number of balls. Therefore, probability that Jason will pick out a red ball from the basket is 4 out of 17 or $\frac{4}{3+4+10} = \frac{4}{17}$
Your Turn!	1) A number is chosen at random from 1 to 20. Find the probability of selecting a prime number. (A prime number is a whole number that is only divisible by itself and 1) _____ 2) There are only red and blue cards in a box. The probability of choosing a red card in the box at random is one third. If there are 24 blue cards, how many cards are in the box? _____ 3) A die is rolled, what is the probability that an even number is obtained? _____

Name: ....   **Date:** ....

Topic	Probability Problems - Answers
Notes	✓ Probability is the likelihood of something happening in the future. It is expressed as a number between zero (can never happen) to 1 (will always happen). ✓ Probability can be expressed as a fraction, a decimal, or a percent. ✓ Probability formula: $Probability = \dfrac{number\ of\ desired\ outcomes}{number\ of\ total\ outcomes}$
Example	***If there are 3 green balls, 4 red balls, and 10 blue balls in a basket, what is the probability that Jason will pick out a red ball from the basket?*** **Solution:** There are 4 red ball and 17 are total number of balls. Therefore, probability that Jason will pick out a red ball from the basket is 4 out of 17 or $\dfrac{4}{3+4+10} = \dfrac{4}{17}$
Your Turn!	1) A number is chosen at random from 1 to 20. Find the probability of selecting a prime number. (A prime number is a whole number that is only divisible by itself and 1) $\dfrac{8}{20} = \dfrac{2}{5}$ *(There are 8 prime numbers from 1 to 20: 2, 3, 5, 7, 11, 13, 17, 19)*
	2) There are only red and blue cards in a box. The probability of choosing a red card in the box at random is one third. If there are 24 blue cards, how many cards are in the box? 36
	3) A die is rolled, what is the probability that an even number is obtained? $\dfrac{1}{2}$

Name:	Date: ....

Topic	**Pie Graph**
Notes	✓ A Pie Chart is a circle chart divided into sectors, each sector represents the relative size of each value.
Example	A library has 460 books that include Mathematics, Physics, Chemistry, English and History. Use following graph to answer the question. **What is the number of Physics books?** **Solution:** Number of total books $= 460$ Percent of Physics books $= 25\% = 0.25$ Then, umber of Physics books: $$0.25 \times 460 = 115$$
Your Turn!	The circle graph below shows all Mr. Smith's expenses for last month. Mr. Smith spent \$440 for clothes last month. Mr. Smith's last month expenses
	1) How much did Mr. Smith spend for his Books last month? _____ 2) How much did Mr. Smith spend for Bills last month? _____ 3) How much did Mr. Smith spend for his foods last month? _____

Name:	**Date:** ....

Topic	**Pie Graph**
Notes	✓ A Pie Chart is a circle chart divided into sectors, each sector represents the relative size of each value.
Example	A library has 460 books that include Mathematics, Physics, Chemistry, English and History. Use following graph to answer the question. **What is the number of Physics books?** **Solution:** Number of total books $= 460$ Percent of Physics books $= 25\% = 0.25$ Then, umber of Physics books: $$0.25 \times 460 = 115$$
Your Turn!	The circle graph below shows all Mr. Smith's expenses for last month. Mr. Smith spent $440 for clothes last month. Mr. Smith's last month expenses
	1) How much did Mr. Smith spend for his Books last month? $308
	2) How much did Mr. Smith spend for Bills last month? $396
	3) How much did Mr. Smith spend for his foods last month? $550

Name:	Date: ....

Topic	**Permutations and Combinations**
Notes	✓ Permutations: The number of ways to choose a sample of k elements from a set of n distinct objects where order does matter, and replacements are not allowed. For a permutation problem, use this formula: $$_nP_k = \frac{n!}{(n-k)!}$$ ✓ Combination: The number of ways to choose a sample of r elements from a set of n distinct objects where order does not matter, and replacements are not allowed. For a combination problem, use this formula: $$_nC_r = \frac{n!}{r!\,(n-r)!}$$ ✓ Factorials are products, indicated by an exclamation mark. For example, 4! Equals: $4 \times 3 \times 2 \times 1$. Remember that 0! is defined to be equal to 1.
Example	***How many ways can we pick a team of 4 people from a group of 8?*** **Solution:** Since the order doesn't matter, we need to use combination formula where n is 8 and r is 4. Then: $\frac{n!}{r!\,(n-r)!} = \frac{8!}{4!\,(8-4)!} = \frac{8!}{4!\,(4)!} = \frac{8\times7\times6\times5\times4!}{4!\,(4)!} = \frac{8\times7\times6\times5}{4\times3\times2\times1} = \frac{1{,}680}{24} = 70$
Your Turn!	1) In how many ways can 8 athletes be arranged in a straight line? _____
	2) How many ways can we award a first and second place prize among eight contestants? _____
	3) In how many ways can we choose 3 players from a team of 9 players? _____

Name:	Date: ....

Topic	**Permutations and Combinations - Answers**
Notes	✓ Permutations: The number of ways to choose a sample of k elements from a set of n distinct objects where order does matter, and replacements are not allowed. For a permutation problem, use this formula: $$_nP_k = \frac{n!}{(n-k)!}$$ ✓ Combination: The number of ways to choose a sample of r elements from a set of n distinct objects where order does not matter, and replacements are not allowed. For a combination problem, use this formula: $$_nC_r = \frac{n!}{r!\,(n-r)!}$$ ✓ Factorials are products, indicated by an exclamation mark. For example, 4! Equals: $4 \times 3 \times 2 \times 1$. Remember that 0! is defined to be equal to 1.
Example	***How many ways can we pick a team of 4 people from a group of 8?*** **Solution:** Since the order doesn't matter, we need to use combination formula where n is 8 and r is 4. Then: $\frac{n!}{r!\,(n-r)!} = \frac{8!}{4!\,(8-4)!} = \frac{8!}{4!\,(4)!} = \frac{8\times7\times6\times5\times4!}{4!\,(4)!} = \frac{8\times7\times6\times5}{4\times3\times2\times1} = \frac{1,680}{24} = 70$
Your Turn!	1) In how many ways can 8 athletes be arranged in a straight line? 40,320
	2) How many ways can we award a first and second place prize among eight contestants? 56
	3) In how many ways can we choose 3 players from a team of 9 players? 84

| Name: .... | Date: .... |

Topic	**Function Notation and Evaluation**
Notes	✓ Functions are mathematical operations that assign unique outputs to given inputs. ✓ Function notation is the way a function is written. It is meant to be a precise way of giving information about the function without a rather lengthy written explanation. ✓ The most popular function notation is $f(x)$ which is read "f of x". ✓ To evaluate a function, plug in the input (the given value or expression) for the function's variable (place holder, x).
Example	**Evaluate**: $h(n) = 2n^2 - 2$, find $h(2)$. **Solution:** Substitute n with 2: Then: $h(n) = 2n^2 - 2 \rightarrow h(2) = 2(2)^2 - 2 = 8 - 2 \rightarrow h(2) = 6$

Your Turn!	1) $f(x) = x - 2$, find $f(-1)$ _____	2) $g(x) = 2x + 4$, find $g(3)$ _____
	3) $g(n) = 2n - 8$, find $g(-1)$ _____	4) $h(n) = n^2 - 1$, find $h(-2)$ _____
	5) $f(x) = x^2 + 12$, find $f(5)$ _____	6) $g(x) = 2x^2 - 9$, find $g(-2)$ _____
	7) $w(x) = 2x^2 - 4x$, find $w(2n)$ _____	8) $p(x) = 4x^3 - 10$, find $p(-3a)$ _____

Name:	Date: ....

Topic	**Function Notation and Evaluation - Answers**	
Notes	✓ Functions are mathematical operations that assign unique outputs to given inputs. ✓ Function notation is the way a function is written. It is meant to be a precise way of giving information about the function without a rather lengthy written explanation. ✓ The most popular function notation is $f(x)$ which is read "f of x". ✓ To evaluate a function, plug in the input (the given value or expression) for the function's variable (place holder, x).	
Example	**Evaluate**: $h(n) = 2n^2 - 2$, find $h(2)$. **Solution:** Substitute n with 2: Then: $h(n) = 2n^2 - 2 \rightarrow h(2) = 2(2)^2 - 2 = 8 - 2 \rightarrow h(2) = 6$	
Your Turn!	1) $f(x) = x - 2$, find $f(-1)$ $f(-1) = -3$	2) $g(x) = 2x + 4$, find $g(3)$ $g(3) = 10$
	3) $g(n) = 2n - 8$, find $g(-1)$ $g(-1) = -10$	4) $h(n) = n^2 - 1$, find $h(-2)$ $h(-2) = 3$
	5) $f(x) = x^2 + 12$, find $f(5)$ $f(5) = 37$	6) $g(x) = 2x^2 - 9$, find $g(-2)$ $g(-2) = -1$
	7) $w(x) = 2x^2 - 4x$, find $w(2n)$ $w(2n) = 8n^2 - 8n$	8) $p(x) = 4x^3 - 10$, find $p(-3a)$ $p(-3a) = -108a^3 + 30a$

Name:	Date: ....

Topic	**Adding and Subtracting Functions**
Notes	✓ Just like we can add and subtract numbers and expressions, we can add or subtract two functions and simplify or evaluate them. The result is a new function. ✓ For two functions $f(x)$ and $g(x)$, we can create two new functions: $(f + g)(x) = f(x) + g(x)$ and $(f - g)(x) = f(x) - g(x)$
Example	$g(a) = 2a - 5, f(a) = a + 8,$ **Find:** $(g + f)(a)$ **Solution:** $(g + f)(a) = g(a) + f(a)$ Then: $(g + f)(a) = (2a - 5) + (a + 8) = 3a + 3$
Your Turn!	1) $g(x) = x - 2$ $h(x) = 2x + 6$ Find: $(h + g)(3)$ _____	2) $f(x) = 3x + 2$ $g(x) = -x - 6$ Find: $(f + g)(2)$ _____
	3) $f(x) = 5x + 8$ $g(x) = 3x - 12$ Find: $(f - g)(-2)$ _____	4) $h(x) = 2x^2 - 10$ $g(x) = 3x + 12$ Find: $(h + g)(3)$ _____
	5) $g(x) = 12x - 8$ $h(x) = 3x^2 + 14$ Find: $(h - g)(x)$ _____	6) $h(x) = -2x^2 - 18$ $g(x) = 4x^2 + 15$ Find: $(h - g)(a)$ _____

| Name: .... | Date: .... |

Topic	**Adding and Subtracting Functions - Answers**
Notes	✓ Just like we can add and subtract numbers and expressions, we can add or subtract two functions and simplify or evaluate them. The result is a new function. ✓ For two functions $f(x)$ and $g(x)$, we can create two new functions: $(f + g)(x) = f(x) + g(x)$ and $(f - g)(x) = f(x) - g(x)$
Example	$g(a) = 2a - 5, f(a) = a + 8$, Find: $(g + f)(a)$ **Solution:** $(g + f)(a) = g(a) + f(a)$ Then: $(g + f)(a) = (2a - 5) + (a + 8) = 3a + 3$

Your Turn!

1) $g(x) = x - 2$

$h(x) = 2x + 6$

Find: $(h + g)(3)$

13

2) $f(x) = 3x + 2$

$g(x) = -x - 6$

Find: $(f + g)(2)$

0

3) $f(x) = 5x + 8$

$g(x) = 3x - 12$

Find: $(f - g)(-2)$

16

4) $h(x) = 2x^2 - 10$

$g(x) = 3x + 12$

Find: $(h + g)(3)$

29

5) $g(x) = 12x - 8$

$h(x) = 3x^2 + 14$

Find: $(h - g)(x)$

$3x^2 - 12x + 22$

6) $h(x) = -2x^2 - 18$

$g(x) = 4x^2 + 15$

Find: $(h - g)(a)$

$-6a^2 - 33$

Name:	Date: ....

Topic	**Multiplying and Dividing Functions**	
Notes	✓ Just like we can multiply and divide numbers and expressions, we can multiply and divide two functions and simplify or evaluate them. ✓ For two functions $f(x)$ and $g(x)$, we can create two new functions: $(f \cdot g)(x) = f(x) \cdot g(x)$ and $\left(\dfrac{f}{g}\right)(x) = \dfrac{f(x)}{g(x)}$	
Example	$g(x) = x + 5, f(x) = x - 3$, Find: $(g \cdot f)(2)$ **Solution:** $(g \cdot f)(x) = g(x) \cdot f(x) = (x + 5)(x - 3) = x^2 - 3x + 5x - 15 = x^2 + 2x - 15$ Substitute x with 2: $(g \cdot f)(x) = (2)^2 + 2(2) - 15 = 4 + 4 - 15 = -7$	
Your Turn!	1) $g(x) = x - 5$ $h(x) = x + 6$ Find: $(g \cdot h)(-1)$ _____	2) $f(x) = 2x + 2$ $g(x) = -x - 6$ Find: $\left(\dfrac{f}{g}\right)(-2)$ _____
	3) $f(x) = 5x + 3$ $g(x) = 2x - 4$ Find: $\left(\dfrac{f}{g}\right)(5)$ _____	4) $h(x) = x^2 - 2$ $g(x) = x + 4$ Find: $(g \cdot h)(3)$ _____
	5) $g(x) = 4x - 12$ $h(x) = x^2 + 4$ Find: $(g \cdot h)(-2)$ _____	6) $h(x) = 3x^2 - 8$ $g(x) = 4x + 6$ Find: $\left(\dfrac{f}{g}\right)(-4)$ _____

Name:	Date: ....

Topic	**Multiplying and Dividing Functions - Answers**
Notes	✓ Just like we can multiply and divide numbers and expressions, we can multiply and divide two functions and simplify or evaluate them. ✓ For two functions $f(x)$ and $g(x)$, we can create two new functions: $(f.g)(x) = f(x).g(x)$ and $\left(\frac{f}{g}\right)(x) = \frac{f(x)}{g(x)}$
Example	$g(x) = x + 5, f(x) = x - 3$, Find: $(g.f)(2)$ **Solution:** $(g.f)(x) = g(x).f(x) = (x + 5)(x - 3) = x^2 - 3x + 5x - 15 = x^2 + 2x - 15$ Substitute x with 2: $(g.f)(x) = (2)^2 + 2(2) - 15 = 4 + 4 - 15 = -7$

Your Turn!	1) $g(x) = x - 5$ $h(x) = x + 6$ Find: $(g.h)(-1)$ $(g.h)(-1) = -30$	2) $f(x) = 2x + 2$ $g(x) = -x - 6$ Find: $\left(\frac{f}{g}\right)(-2)$ $\left(\frac{f}{g}\right)(-2) = \frac{1}{2}$
	3) $f(x) = 5x + 3$ $g(x) = 2x - 4$ Find: $\left(\frac{f}{g}\right)(5)$ $\left(\frac{f}{g}\right)(5) = \frac{14}{3}$	4) $h(x) = x^2 - 2$ $g(x) = x + 4$ Find: $(g.h)(3)$ $(g.h)(3) = 49$
	5) $g(x) = 4x - 12$ $h(x) = x^2 + 4$ Find: $(g.h)(-2)$ $(g.h)(-2) = -160$	6) $h(x) = 3x^2 - 8$ $g(x) = 4x + 6$ Find: $\left(\frac{f}{g}\right)(-4)$ $\left(\frac{f}{g}\right)(-4) = -4$

Name:	Date: ....

Topic	**Composition of Functions**	
Notes	✓ "Composition of functions" simply means combining two or more functions in a way where the output from one function becomes the input for the next function. ✓ The notation used for composition is: $(fog)(x) = f(g(x))$ and is read "f composed with g of x" or "f of g of x".	
Example	*Using* $f(x) = x - 8$ *and* $g(x) = x + 2$, *find:* $(f \circ g)(3)$ **Solution:** $(f \circ g)(x) = f(g(x))$ *Then:* $(f \circ g)(x) = f(g(x)) = f(x + 2) = x + 2 - 8 = x - 6$ Substitute x with 3: $(f \circ g)(3) = f(g(3)) = 3 - 6 = -3$	
Your Turn!	1) $f(x) = 2x$ $g(x) = x + 3$ Find: $(fog)(2)$ _____	2) $f(x) = x + 2$ $g(x) = x - 6$ Find: $(fog)(-1)$ _____
	3) $f(x) = 3x$ $g(x) = x + 4$ Find: $(gof)(4)$ _____	4) $h(x) = 2x - 2$ $g(x) = x + 4$ Find: $(goh)(2)$ _____
	5) $f(x) = 2x - 8$ $g(x) = x + 10$ Find: $(fog)(-2)$ _____	6) $f(x) = x^2 - 8$ $g(x) = 2x + 3$ Find: $(gof)(4)$ _____

| Name: .... | Date: .... |

Topic	Composition of Functions - Answers
Notes	✓ "Composition of functions" simply means combining two or more functions in a way where the output from one function becomes the input for the next function. ✓ The notation used for composition is: $(fog)(x) = f(g(x))$ and is read "f composed with g of x" or "f of g of x".
Example	*Using* $f(x) = x - 8$ *and* $g(x) = x + 2$, *find:* $(f \circ g)(3)$ **Solution:** $(f \circ g)(x) = f(g(x))$ *Then:* $(f \circ g)(x) = f(g(x)) = f(x + 2) = x + 2 - 8 = x - 6$ Substitute x with 3: $(f \circ g)(3) = f(g(3)) = 3 - 6 = -3$

Your Turn!	1) $f(x) = 2x$ $g(x) = x + 3$ Find: $(fog)(2)$ 10	2) $f(x) = x + 2$ $g(x) = x - 6$ Find: $(fog)(-1)$ -5
	3) $f(x) = 3x$ $g(x) = x + 4$ Find: $(gof)(4)$ 16	4) $h(x) = 2x - 2$ $g(x) = x + 4$ Find: $(goh)(2)$ 6
	5) $f(x) = 2x - 8$ $g(x) = x + 10$ Find: $(fog)(-2)$ 8	6) $f(x) = x^2 - 8$ $g(x) = 2x + 3$ Find: $(gof)(4)$ 19

Name: ....   Date: ....

Topic	**Solving a Quadratic Equation**
Notes	✓ Write the equation in the form of: $ax^2 + bx + c = 0$ ✓ Factor the quadratic and solve for the variable. ✓ Use quadratic formula if you couldn't factorize the quadratic. ✓ Quadratic formula: $x = \dfrac{-b \pm \sqrt{b^2 - 4ac}}{2a}$
Example	**Find the solutions of quadratic.** $x^2 + x - 72 = 0$ **Solution:** Use quadratic formula: $x = \dfrac{-b \pm \sqrt{b^2 - 4ac}}{2a}$, $a = 1, b = 1$ and $c = -72$ $x = \dfrac{-1 \pm \sqrt{1^2 - 4 \times 1(-72)}}{2 \times 1}$ $x_1 = \dfrac{-1 + \sqrt{1^2 - 4 \times 1 \times (-72)}}{2 \times 1} = 8$, $x_2 = \dfrac{-1 - \sqrt{1^2 - 4 \times 1 \times (-72)}}{2 \times 1} = -9$
Your Turn!	1) $x^2 - x - 2 = 0$ $x = \_\_\_$, $x = \_\_\_$ 2) $x^2 - 6x + 8 = 0$ $x = \_\_\_$, $x = \_\_\_$ 3) $x^2 - 4x + 3 = 0$ $x = \_\_\_$, $x = \_\_\_$ 4) $x^2 + x - 12 = 0$ $x = \_\_\_$, $x = \_\_\_$ 5) $x^2 + 7x - 18 = 0$ $x = \_\_\_$, $x = \_\_\_$ 6) $x^2 - 2x - 15 = 0$ $x = \_\_\_$, $x = \_\_\_$ 7) $x^2 + 6x - 40 = 0$ $x = \_\_\_$, $x = \_\_\_$ 8) $x^2 - 9x - 36 = 0$ $x = \_\_\_$, $x = \_\_\_$

Name: ....

Date: ....

Topic	**Solving a Quadratic Equation- Answers**
Notes	✓ Write the equation in the form of: $ax^2 + bx + c = 0$ ✓ Factor the quadratic and solve for the variable. ✓ Use quadratic formula if you couldn't factorize the quadratic. ✓ Quadratic formula: $x = \dfrac{-b \pm \sqrt{b^2 - 4ac}}{2a}$
Example	**Find the solutions of quadratic.** $x^2 + x - 72 = 0$ **Solution:** Use quadratic formula: $x = \dfrac{-b \pm \sqrt{b^2 - 4ac}}{2a}$, $a = 1, b = 1$ and $c = -72$ $x = \dfrac{-1 \pm \sqrt{1^2 - 4 \times 1(-72)}}{2 \times 1}$ $x_1 = \dfrac{-1 + \sqrt{1^2 - 4 \times 1 \times (-72)}}{2 \times 1} = 8$, $x_2 = \dfrac{-1 - \sqrt{1^2 - 4 \times 1 \times (-72)}}{2 \times 1} = -9$
Your Turn!	1) $x^2 - x - 2 = 0$ $x = 2, x = -1$ 2) $x^2 - 6x + 8 = 0$ $x = 2, x = 4$ 3) $x^2 - 4x + 3 = 0$ $x = 3, x = 1$ 4) $x^2 + x - 12 = 0$ $x = 3, x = -4$ 5) $x^2 + 7x - 18 = 0$ $x = 2, x = -9$ 6) $x^2 - 2x - 15 = 0$ $x = 5, x = -3$ 7) $x^2 + 6x - 40 = 0$ $x = 4, x = -10$ 8) $x^2 - 9x - 36 = 0$ $x = 12, x = -3$

Name: ....   **Date:** ....

Topic	**Graphing Quadratic Functions**
Notes	✓ Quadratic functions in vertex form: $y = a(x - h)^2 + k$ where (h, k) is the vertex of the function. The axis of symmetry is $x = h$ ✓ Quadratic functions in standard form: $y = ax^2 + bx + c$ where $x = -\dfrac{b}{2a}$ is the value of x in the vertex of the function. ✓ To graph a quadratic function, first find the vertex, then substitute some values for x and solve for y.
Example	***Sketch the graph of*** $y = (x - 2)^2 - 5$ **Solution:** *The vertex of* $y = (x - 2)^2 - 5$ *is* $(2, 5)$. Substitute zero for x and solve for y. $$y = (0 - 2)^2 - 5 = -1$$ The y-_Intercept_ is $(0, -1)$ Now, you can simply graph the quadratic function.
Your Turn!	1) $y = (x - 4)^2 - 2$ 2) $y = 2(x + 2)^2 - 3$

Name: ....

Date: ....

Topic	Graphing Quadratic Functions- Answers
Notes	✓ Quadratic functions in vertex form: $y = a(x - h)^2 + k$ where (h, k) is the vertex of the function. The axis of symmetry is $x = h$ ✓ Quadratic functions in standard form: $y = ax^2 + bx + c$ where $x = -\frac{b}{2a}$ is the value of x in the vertex of the function. ✓ To graph a quadratic function, first find the vertex, then substitute some values for x and solve for y.
Example	**Sketch the graph of** $y = (x - 2)^2 - 5$ **Solution:** *The vertex of $y = (x - 2)^2 - 5$ is $(2, 5)$.* Substitute zero for x and solve for y. $\quad y = (0 - 2)^2 - 5 = -1$ The y-*Intercept* is $(0, -1)$ Now, you can simply graph the quadratic function.
Your Turn!	1) $y = (x - 4)^2 - 2$ 2) $y = 2(x + 2)^2 - 3$

Name: ....   Date: ....

Topic	Solving Quadratic Inequalities
Notes	✓ A quadratic inequality is one that can be written in the standard form of $ax^2 + bx + c > 0$ (or substitute $<$, $\leq$, or $\geq$ for $>$). ✓ Solving a quadratic inequality is like solving equations. We need to find the solutions (the zeroes). ✓ To solve quadratic inequalities, first find quadratic equations. Then choose a test value between zeroes. Finally, find interval(s), such as > 0 or < 0.
Example	**Solve quadratic inequality.** $x^2 + x - 12 > 0$ **Solution:** First solve $x^2 + x - 12 = 0$ by factoring. Then: $x^2 + x - 6 = 0 \rightarrow$ $(x - 3)(x + 4) = 0$. The product of two expressions is 0. Then: $(x - 3) = 0 \rightarrow x = 3$ or $(x + 4) = 0 \rightarrow x = -4$. Now, choose a value between 3 and -4. Let's choose 0. Then: $x = 0 \rightarrow x^2 + x - 12 > 0 \rightarrow (0)^2 + (0) - 12 > 0 \rightarrow -12 > 0$ -12 is not greater than 0. Therefore, all values between 3 and -4 are NOT the solution of this quadratic inequality. The solution is: $x > 3$ and $x < -4$.
Your Turn!	1) $x^2 - 6x - 27 > 0$ _____ 2) $x^2 + 13x + 42 < 0$ _____ 3) $x^2 + x - 56 > 0$ _____ 4) $x^2 - 15x + 54 < 0$ _____ 5) $x^2 + 2x - 35 \leq 0$ _____ 6) $x^2 - x - 72 \geq 0$ _____

Name: ....   **Date:** ....

Topic	Solving Quadratic Inequalities
Notes	✓ A quadratic inequality is one that can be written in the standard form of $ax^2 + bx + c > 0$ (or substitute $<, \leq,$ or $\geq$ for $>$). ✓ Solving a quadratic inequality is like solving equations. We need to find the solutions (the zeroes). ✓ To solve quadratic inequalities, first find quadratic equations. Then choose a test value between zeroes. Finally, find interval(s), such as > 0 or < 0.
Example	**Solve quadratic inequality.** $x^2 + x - 12 > 0$ **Solution:** First solve $x^2 + x - 12 = 0$ by factoring. Then: $x^2 + x - 6 = 0 \rightarrow$ $(x - 3)(x + 4) = 0$. The product of two expressions is 0. Then: $(x - 3) = 0 \rightarrow x = 3$ or $(x + 4) = 0 \rightarrow x = -4$. Now, choose a value between 3 and -4. Let's choose 0. Then: $x = 0 \rightarrow x^2 + x - 12 > 0 \rightarrow (0)^2 + (0) - 12 > 0 \rightarrow -12 > 0$ -12 is not greater than 0. Therefore, all values between 3 and -4 are NOT the solution of this quadratic inequality. The solution is: $x > 3$ and $x < -4$.

Your Turn!	1) $x^2 - 6x - 27 > 0$ $x < -3 \; or \; x > 9$	2) $x^2 + 13x + 42 < 0$ $-7 < x < -6$
	3) $x^2 + x - 56 > 0$ $x < -8 \; or \; x > 7$	4) $x^2 - 15x + 54 < 0$ $6 < x < 9$
	5) $x^2 + 2x - 35 \leq 0$ $-7 \leq x \leq 5$	6) $x^2 - x - 72 \geq 0$ $x \leq -8 \; or \; x \geq 9$

Name: .... **Date:** ....

Topic	**Graphing Quadratic Inequalities**
Notes	✓ A quadratic inequality is in the form $y > ax^2 + bx + c$ (or substitute $<, \leq,$ or $\geq$ for $>$). ✓ To graph a quadratic inequality, start by graphing the quadratic parabola. Then fill in the region either inside or outside of it, depending on the inequality. ✓ Choose a testing point and check the solution section.
Example	**Sketch the graph of $y > x^2$** **Solution:** First, graph $y = x^2$ Since, the inequality sing is $>$, we need to use dash lines. Now, choose a testing point inside the parabola. Let's choose $(0,2)$. $y > x^2 \rightarrow 2 > (0)^2 \rightarrow 2 > 0$ This is true. So, inside the parabola is the solution section.
Your Turn!	1) $y \leq x^2 + 4x + 5$ 2) $y \leq x^2 + 2x - 3$

Name: ....   Date: ....

Topic	Graphing Quadratic Inequalities- Answers
Notes	✓ A quadratic inequality is in the form $y > ax^2 + bx + c$ (or substitute $<$, $\leq$, or $\geq$ for $>$). ✓ To graph a quadratic inequality, start by graphing the quadratic parabola. Then fill in the region either inside or outside of it, depending on the inequality. ✓ Choose a testing point and check the solution section.
Example	**Sketch the graph of $y > x^2$** **Solution:** First, graph $y = x^2$ Since, the inequality sing is $>$, we need to use dash lines. Now, choose a testing point inside the parabola. Let's choose $(0,2)$. $y > x^2 \rightarrow 2 > (0)^2 \rightarrow 2 > 0$ This is true. So, inside the parabola is the solution section.
Your Turn!	1) $y \leq x^2 + 4x + 5$ 2) $y \leq x^2 + 2x - 3$

Time to test

Time to refine your skill with a practice examination

In this section, there are two complete PERT Mathematics Tests. Take these tests to simulate the test day experience. After you've finished, score your test using the answer key.

Before You Start

- You'll need a pencil, a timer, and a four-function calculator to take the test.

- Use the answer sheet provided to record your answers. (You can cut it out or photocopy it)

- For each question there are four possible answers. Choose which one is best.

- After you've finished the test, review the answer key to see where you went wrong and what areas you need to improve.

Good luck!

PERT Math Practice Test 1

2021 - 2022

Total number of questions: 30

Total time: No time limit

You may use a calculator on this test.

PERT Mathematics Practice Test Answer Sheet

Remove (or photocopy) this answer sheet and use it to complete the practice test.

PERT Mathematics Practice Test 1 Answer Sheet		
1 Ⓐ Ⓑ Ⓒ Ⓓ	13 Ⓐ Ⓑ Ⓒ Ⓓ	25 Ⓐ Ⓑ Ⓒ Ⓓ
2 Ⓐ Ⓑ Ⓒ Ⓓ	14 Ⓐ Ⓑ Ⓒ Ⓓ	26 Ⓐ Ⓑ Ⓒ Ⓓ
3 Ⓐ Ⓑ Ⓒ Ⓓ	15 Ⓐ Ⓑ Ⓒ Ⓓ	27 Ⓐ Ⓑ Ⓒ Ⓓ
4 Ⓐ Ⓑ Ⓒ Ⓓ	16 Ⓐ Ⓑ Ⓒ Ⓓ	28 Ⓐ Ⓑ Ⓒ Ⓓ
5 Ⓐ Ⓑ Ⓒ Ⓓ	17 Ⓐ Ⓑ Ⓒ Ⓓ	29 Ⓐ Ⓑ Ⓒ Ⓓ
6 Ⓐ Ⓑ Ⓒ Ⓓ	18 Ⓐ Ⓑ Ⓒ Ⓓ	30 Ⓐ Ⓑ Ⓒ Ⓓ
7 Ⓐ Ⓑ Ⓒ Ⓓ	19 Ⓐ Ⓑ Ⓒ Ⓓ	
8 Ⓐ Ⓑ Ⓒ Ⓓ	20 Ⓐ Ⓑ Ⓒ Ⓓ	
9 Ⓐ Ⓑ Ⓒ Ⓓ	21 Ⓐ Ⓑ Ⓒ Ⓓ	
10 Ⓐ Ⓑ Ⓒ Ⓓ	22 Ⓐ Ⓑ Ⓒ Ⓓ	
11 Ⓐ Ⓑ Ⓒ Ⓓ	23 Ⓐ Ⓑ Ⓒ Ⓓ	
12 Ⓐ Ⓑ Ⓒ Ⓓ	24 Ⓐ Ⓑ Ⓒ Ⓓ	

1) If $3x - 5 = 8.5$, what is the value of $6x + 3$?

   ☐A. 13   ☐B. 15.5

   ☐C. 20.5   ☐D. 30

2) What is the area of an isosceles right triangle that has one leg that measures $8\ cm$?

   ☐A. $6\ cm^2$ ☐B. $12\ cm^2$

   ☐C. $18\ cm^2$ ☐D. $32\ cm^2$

3) A shirt costing \$600 is discounted 25%. After a month, the shirt is discounted another 15%. Which of the following expressions can be used to find the selling price of the shirt?

   ☐A. $(600)(0.60)$ ☐B. $(600) - 600\ (0.40)$

   ☐C. $(600)(0.25) - (200)(0.15)$ ☐D. $(600)(0.75)(0.85)$

4) Which of the following points lies on the line with equation $3x + 5y = 11$?

   ☐A. $(2, 1)$ ☐B. $(-1, 2)$

   ☐C. $(-2, 2)$ ☐D. $(2, 2)$

5) The average of five consecutive numbers is 40. What is the smallest number?

   ☐A. 38   ☐B. 36

   ☐C. 34   ☐D. 12

6) How many tiles of $8\ cm^2$ is needed to cover a floor of dimension $7\ cm$ by $24\ cm$?

   ☐A. 6   ☐B. 12

   ☒C. 21   ☐D. 24

7) A rope weighs 600 grams per meter of length. What is the weight in kilograms of 15.2 meters of this rope? ($1\ kilograms = 1,000\ grams$)

    ☐A. 0.0912   ☐B. 0.912

    ☐C. 9.12   ☐D. 91.20

8) A chemical solution contains 6% alcohol. If there is $24\ ml$ of alcohol, what is the volume of the solution?

    ☐A. $240\ ml$ ☐B. $400\ ml$

    ☐C. $600\ ml$ ☐D. $1200\ ml$

9) The average weight of 18 girls in a class is $60\ kg$ and the average weight of 32 boys in the same class is $62\ kg$. What is the average weight of all the 50 students in that class?

    ☐A. 60   ☐B. 61.28

    ☐C. 61.68   ☐D. 62.90

10) The price of a laptop is decreased by 10% to $360. What is its original price?

    ☐A. $320   ☐B. $380

    ☐C. $400   ☐D. $450

11) What is the median of these numbers? 4, 9, 13, 8, 15, 18, 5

    ☐A. 8   ☐B. 9

    ☐C. 13   ☐D. 15

12) In 1999, the average worker's income increased $2,000 per year starting from $27,000 annual salary. Which equation represents income greater than average? (I = income, x = number of years after 1999)

☐A. $I > 2,000 x + 27,000$

☐B. $I > -2,000 x + 27,000$

☐C. $I < -2,000 x + 27,000$

☐D. $I < 2,000 x - 27,000$

13) What is the value of y in the following system of equation?

$$3x - 4y = -20$$

$$-x + 2y = 10$$

☐A. 2

☐B. 4

☐C. 5

☐D. 8

14) What is the area of a square whose diagonal is 6 meters?

☐A. $20\ m^2$

☐C. $12\ m^2$

☐B. $18\ m^2$

☐D. $10\ m^2$

15) The width of a box is one third of its length. The height of the box is half of its width. If the length of the box is 24 cm, what is the volume of the box?

☐A. $81\ cm^3$

☐B. $162\ cm^3$

☐C. $243\ cm^3$

☐D. $768\ cm^3$

16) If 60% of A is 20% of B, then B is what percent of A?

☐A. 3%

☐B. 30%

☐C. 200%

☐D. 300%

17) A bank is offering 2.5% simple interest on a savings account. If you deposit $8,000, how much interest will you earn in five years?

  ☐A. $360   ☐B. $720

  ☐C. $1,000   ☐D. $3,600

18) 15 is What percent of 20?

  ☐A. 20%   ☐B. 75%

  ☐C. 125%   ☐D. 150%

19) In five successive hours, a car travels $40\ km$, $45\ km$, $50\ km$, $35\ km$ and $55\ km$. In the next five hours, it travels with an average speed of $45\ km\ per\ hour$. Find the total distance the car traveled in $10\ hours$.

  ☐A. $425\ km$ ☐B. $450\ km$

  ☐C. $475\ km$ ☐D. $500\ km$

20) How long does a $420-miles$ trip take moving at $60\ miles\ per\ hour\ (mph)$?

  ☐A. $4\ hours$ ☐B. $7\ hours$

  ☐C. $7\ hours\ and\ 24\ minutes$ ☐D. $8\ hours\ and\ 10\ minutes$

21) Which of the following points lies on the line $4x + 6y = 20$?

  ☐A. $(2, 1)$ ☐B. $(-1, 3)$

  ☐C. $(-2, 2)$ ☐D. $(2, 2)$

22) Two third of 15 is equal to $\frac{2}{5}$ of what number?

  ☐A. 12   ☐B. 20

  ☐C. 25   ☐D. 60

23) The marked price of a computer is D dollar. Its price decreased by 20% in January and later increased by 15% in February. What is the final price of the computer in D dollar?

☐A. 0.80 D ☐B. 0.88 D

☐C. 0.92 D ☐D. 1.20

24) A $45 shirt now selling for $28 is discounted by about what percent?

☐A. 20%   ☐B. 37.7%

☐C. 40%   ☐D. 60%

25) How many meters is 27,356 centimeters?

☐ A. 27.356 m

☐ C. 2.735600 m

☐ B. 273.5600 m

☒ D. 2,735.600 m

26) The score of Emma was half as that of Ava and the score of Mia was twice that of Ava. If the score of Mia was 60, what is the score of Emma?

☐A. 12   ☐B. 15

☐C. 20   ☐D. 30

27) A bag contains 21 balls: two green, six black, eight blue, two brown, two red and one white. If 20 balls are removed from the bag at random, what is the probability that a white ball has been removed?

☐A. $\dfrac{1}{9}$ ☐B. $\dfrac{1}{6}$

☐C. $\dfrac{4}{5}$ ☐D. $\dfrac{20}{21}$

28) A taxi driver earns $8 per 1-hour work. If he works 10 hours a day and in 1 hour he uses 2-liters petrol with price $1 for 1-liter. How much money does he earn in one day?

   ☐A. $90   ☐B. $88

   ☐C. $70   ☐D. $60

29) The price of a sofa is decreased by 15% to $476. What was its original price?

   ☐A. $480   ☐B. $520

   ☐C. $560   ☐D. $600

30) When a number is subtracted from 28 and the difference is divided by that number, the result is 3. What is the value of the number?

   ☐A. 2   ☐B. 4

   ☐C. 7   ☐D. 12

End of PERT Mathematics Practice Test

PERT Math Practice Test 2

2021- 2022

Total number of questions: 30

Total time: No time limit

You may use a calculator on this test.

PERT Mathematics Practice Test Answer Sheet

Remove (or photocopy) this answer sheet and use it to complete the practice test.

PERT Mathematics Practice Test 2 Answer Sheet		
1 Ⓐ Ⓑ Ⓒ Ⓓ	13 Ⓐ Ⓑ Ⓒ Ⓓ	25 Ⓐ Ⓑ Ⓒ Ⓓ
2 Ⓐ Ⓑ Ⓒ Ⓓ	14 Ⓐ Ⓑ Ⓒ Ⓓ	26 Ⓐ Ⓑ Ⓒ Ⓓ
3 Ⓐ Ⓑ Ⓒ Ⓓ	15 Ⓐ Ⓑ Ⓒ Ⓓ	27 Ⓐ Ⓑ Ⓒ Ⓓ
4 Ⓐ Ⓑ Ⓒ Ⓓ	16 Ⓐ Ⓑ Ⓒ Ⓓ	28 Ⓐ Ⓑ Ⓒ Ⓓ
5 Ⓐ Ⓑ Ⓒ Ⓓ	17 Ⓐ Ⓑ Ⓒ Ⓓ	29 Ⓐ Ⓑ Ⓒ Ⓓ
6 Ⓐ Ⓑ Ⓒ Ⓓ	18 Ⓐ Ⓑ Ⓒ Ⓓ	30 Ⓐ Ⓑ Ⓒ Ⓓ
7 Ⓐ Ⓑ Ⓒ Ⓓ	19 Ⓐ Ⓑ Ⓒ Ⓓ	
8 Ⓐ Ⓑ Ⓒ Ⓓ	20 Ⓐ Ⓑ Ⓒ Ⓓ	
9 Ⓐ Ⓑ Ⓒ Ⓓ	21 Ⓐ Ⓑ Ⓒ Ⓓ	
10 Ⓐ Ⓑ Ⓒ Ⓓ	22 Ⓐ Ⓑ Ⓒ Ⓓ	
11 Ⓐ Ⓑ Ⓒ Ⓓ	23 Ⓐ Ⓑ Ⓒ Ⓓ	
12 Ⓐ Ⓑ Ⓒ Ⓓ	24 Ⓐ Ⓑ Ⓒ Ⓓ	

1) The sum of two numbers is x. If one of the numbers is 9, then two times the other number would be?

☐A. $2x$ ☐B. $2 + x \times 2$

☐C. $2(x + 9)$ ☐D. $2(x - 9)$

2) A tree 32 feet tall casts a shadow 12 feet long. Jack is 6 feet tall. How long is Jack's shadow?

☐A. $2.25\ ft$ ☐B. $4\ ft$

☐C. $4.25\ ft$ ☐D. $8\ ft$

3) What is the product of all possible values of x in the following equation?

$$|2x - 6| = 12$$

☐A. -27 ☐B. -3

☐C. 9 ☐D. 27

4) What is the slope of a line that is perpendicular to the line $3x - y = 6$?

☐A. -3 ☐B. $-\frac{1}{3}$

☐C. 2 ☐D. 6

5) What is the value of the expression $3(x - 2y) + (2 - x)^2$ when $x = 5$ and $y = -3$?

☐A. -22 ☐B. 24

☐C. 42 ☐D. 88

6) $\dfrac{(15\ feet + 7\ yards)}{4} = $ _____

☐A. $4\ ft$ ☐B. $7\ ft$

☐C. $9\ ft$ ☐D. $28\ ft$

7) Which of the following answers represents the compound inequality $-4 \le 4x - 8 < 16$?

☐A. $-2 \le x \le 8$ ☐C. $1 < x \le 6$

☐B. $-2 < x \le 8$ ☐D. $1 \le x < 6$

8) What is the volume of a box with the following dimensions?

Hight = $4\ cm$ Width = $5\ cm$ Length = $6\ cm$

☐A. $15\ cm^3$ ☐B. $60\ cm^3$

☐C. $90\ cm^3$ ☐D. $120\ cm^3$

9) Simplify the expression.

$$(6x^3 - 8x^2 + 2x^4) - (4x^2 - 2x^4 + 2x^3)$$

☐A. $4x^4 + 4x^3 - 12x^2$ ☐B. $4x^3 - 12x^2$

☐C. $4x^4 + 4x^3 + 12x^2$ ☐D. $8x^3 - 12x^2$

10) In two successive years, the population of a town is increased by 15% and 20%. What percent of the population is increased after two years?

☐A. 32%   ☐B. 35%

☐C. 38%   ☐D. 68%

11) Last week 24,000 fans attended a football match. This week three times as many bought tickets, but one sixth of them cancelled their tickets. How many are attending this week?

☐A. 48,000 ☐B. 54,000

☐C. 60,000 ☐D. 72,000

12) $\frac{7}{25}$ is equals to:

☐A. 0.3 ☐B. 2.8

☐C. 0.03 ☐D. 0.28

13) In the simplest form, $\frac{18}{24}$ is

☐A. $\frac{2}{3}$ ☐B. $\frac{3}{2}$

☐C. $\frac{4}{3}$ ☐D. $\frac{3}{4}$

14) The mean of 50 test scores was calculated as 88. But, it turned out that one of the scores was misread as 94 but it was 69. What is the correct mean of the test scores?

☐A. 85 ☐B. 87

☐C. 87.5 ☐D. 88.5

15) If two angles in a triangle measure 53 degrees and 45 degrees, what is the value of the third angle?

☐A. 8 *degrees* ☐B. 42 *degrees*

☐C. 82 *degrees* ☐D. 98 *degrees*

16) What is the area of a square whose diagonal is 8?

☐A. 16   ☐B. 32

☐C. 36   ☐D. 64

17) Anita's trick–or–treat bag contains 12 pieces of chocolate, 18 suckers, 18 pieces of gum, 24 pieces of licorice. If she randomly pulls a piece of candy from her bag, what is the probability of her pulling out a piece of sucker?

☐A. $\dfrac{1}{3}$ ☐B. $\dfrac{1}{4}$

☐C. $\dfrac{1}{6}$ ☐D. $\dfrac{1}{12}$

18) The perimeter of a rectangular yard is 60 meters. What is its length if its width is twice its length?

☐A. 10 meters   ☐B. 18 meters

☐C. 20 meters   ☐D. 24 meters

19) The average of 6 numbers is 12. The average of 4 of those numbers is 10. What is the average of the other two numbers?

☐A. 10   ☐B. 12

☐C. 14   ☐D. 16

20) What is the value of x in the following system of equations?

$$2x + 5y = 11$$
$$4x - 2y = -14$$

☐A. -1 ☐B. 1

☐C. -2 ☐D. 4

21) The perimeter of the trapezoid below is $36\ cm$. What is its area?

12 cm

☐A. $576\ cm^2$ ☐B. $70\ cm^2$

☐C. $48\ cm^2$ ☐D. $24\ cm^2$ 6 cm   8 cm

22) A card is drawn at random from a standard 52–card deck, what is the probability that the card is of Hearts? (The deck includes 13 of each suit clubs, diamonds, hearts, and spades)

☐A. $\dfrac{1}{3}$ ☐B. $\dfrac{1}{4}$

☐C. $\dfrac{1}{6}$ ☐D. $\dfrac{1}{52}$

23) The ratio of boys and girls in a class is $4:7$. If there are 44 students in the class, how many more boys should be enrolled to make the ratio $1:1$?

☐A. 8   ☐B. 10

☐C. 12   ☐D. 14

24) Mr. Jones saves $2,500 out of his monthly family income of $55,000. What fractional part of his income does he save?

□A. $\frac{1}{22}$ □B. $\frac{1}{11}$

□C. $\frac{3}{25}$ □D. $\frac{2}{15}$

25) What is the value of x in the following equation? $\frac{2}{3}x + \frac{1}{6} = \frac{1}{3}$

□A. 6   □B. $\frac{1}{2}$

□C. $\frac{1}{3}$ □D. $\frac{1}{4}$

26) A bank is offering 3.5% simple interest on a savings account. If you deposit $12,000, how much interest will you earn in two years?

□A. $420   □B. $840

□C. $4200   □D. $8,400

27) Simplify $6x^2y^3(2x^2y)^3 =$

□A. $12x^4y^6$ □B. $12x^8y^6$

□C. $48x^4y^6$ □D. $48x^8y^6$

28) What is the surface area of the cylinder below?

☐A. $48\,\pi\,in^2$ ☐B. $57\,\pi\,in^2$

☐C. $66\,\pi\,in^2$ ☐D. $288\,\pi\,in^2$

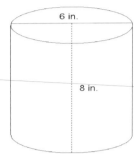

6 in.

8 in.

29) What is the median of these numbers? $2, 27, 28, 19, 67, 44, 35$

☐A. 19   ☐B. 28

☐C. 44   ☐D. 35

30) Which graph corresponds to the following inequality?

$$-6y \leq 16x - 12$$

☐A.

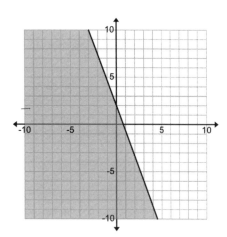

☐B.

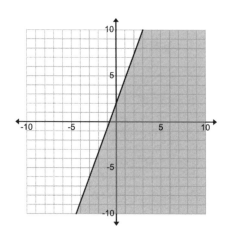

☐C.

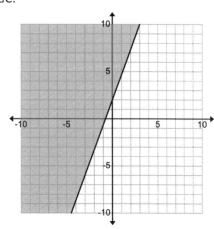

☐D.

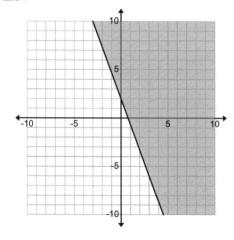

End of PERT Mathematics Practice Test

PERT Mathematics Practice Tests Answer Keys

Now, it's time to review your results to see where you went wrong and what areas you need to improve.

PERT Math Practice Test 1				PERT Math Practice Test 2			
1	D	21	D	1	D	21	B
2	D	22	C	2	A	22	B
3	D	23	C	3	A	23	C
4	A	24	B	4	B	24	A
5	A	25	B	5	C	25	D
6	C	26	B	6	C	26	B
7	C	27	D	7	D	27	D
8	B	28	D	8	D	28	C
9	B	29	C	9	A	29	B
10	C	30	C	10	C	30	D
11	B			11	C		
12	A			12	D		
13	C			13	D		
14	B			14	C		
15	D			15	C		
16	D			16	B		
17	C			17	B		
18	B			18	A		
19	B			19	D		
20	B			20	C		

PERT Mathematics Practice Tests Answers and Explanations

PERT Mathematics Practice Test 1

Answers and Explanations

1) Choice D is correct

$3x - 5 = 8.5 \rightarrow 3x = 8.5 + 5 = 13.5 \rightarrow x = \frac{13.5}{3} = 4.5$

Then; $6x + 3 = 6(4.5) + 3 = 27 + 3 = 30$

2) Choice D is correct

First draw an isosceles triangle. Remember that two sides of the triangle are equal.

     Let put a for the legs. Then:

Isosceles right triangle

$a = 8 \Rightarrow$ area of the triangle is $= \frac{1}{2}(8 \times 8) = \frac{64}{2} = 32 \; cm^2$

3) Choice D is correct

To find the discount, multiply the number by $(100\% - rate \; of \; discount)$.

Therefore, for the first discount we get: $(600)(100\% - 25\%) = (600)(0.75)$

For the next 15% discount: $(600)(0.75)(0.85)$

4) Choice A is correct

Plug in each pair of numbers in the equation: $3x + 5y = 11$

     A. $(2,1):$ $3(2) + 5(1) = 11$
     B. $(-1,2):$ $3(-1) + 5(2) = 7$
     C. $(-2,2):$ $3(-2) + 5(2) = 4$
     D. $(2,2):$ $3(2) + 5(2) = 16$

Choice A is correct.

5) Choice A is correct

Let x be the smallest number. Then, these are the numbers: $x, x + 1, x + 2, x + 3, x + 4$

average $= \frac{\text{sum of terms}}{\text{number of terms}} \Rightarrow 40 = \frac{x+(x+1)+(x+2)+(x+3)+(x+4)}{5} \Rightarrow 40 = \frac{5x+10}{5} \Rightarrow 200 = 5x + 10$

$\Rightarrow 190 = 5x \Rightarrow x = 38$

6) Choice C is correct

The area of the floor is: $7\ cm \times 24\ cm = 168\ cm^2$, The number of tiles needed $=$

$$168 \div 8 = 21$$

7) Choice C is correct

The weight of 15.2 meters of this rope is: $15.2 \times 600\ g = 9{,}120\ g$, $1\ kg = 1{,}000\ g$, therefore, $7{,}320\ g \div 1000 = 9.12\ kg$

8) Choice B is correct

6% of the volume of the solution is alcohol. Let x be the volume of the solution.

Then: $6\%\ of\ x = 24\ ml \Rightarrow 0.06\ x = 24 \Rightarrow x = 24 \div 0.06 = 400$

9) Choice B is correct

$$average\ = \frac{sum\ of\ terms}{number\ of\ terms}$$

The sum of the weight of all girls is: $18 \times 60 = 1080\ kg$, The sum of the weight of all boys is: $32 \times 62 = 1984\ kg$, The sum of the weight of all students is: $1080 + 1984 = 3064\ kg$

$$average\ = \frac{3064}{50} = 61.28$$

10) Choice C is correct

Let x be the original price. If the price of a laptop is decreased by 10% to $360, then:

$90\%\ of\ x = 360 \Rightarrow 0.90x = 360 \Rightarrow x = 360 \div 0.90 = 400$

11) Choice B is correct

Write the numbers in order: $4, 5, 8, 9, 13, 15, 18$

Since we have 7 numbers (7 is odd), then the median is the number in the middle, which is 9.

12) Choice A is correct

Let x be the number of years. Therefore, $\$2{,}000\ per\ year$ equals $2000x$. starting from $27,000 annual salary means you should add that amount to $2000x$.

Income more than that is: $I > 2000x + 27000$

13) Choice C is correct

Solving Systems of Equations by Elimination

$\begin{array}{l} 3x - 4y = -20 \\ -x + 2y = 10 \end{array}$ Multiply the second equation by 3, then add it to the first equation.

$$3x - 4y = -20 \atop 3(-x + 2y = 10)} \Rightarrow {3x - 4y = -20 \atop -3x + 6y = 30} \Rightarrow 2y = 10 \Rightarrow y = 5$$

14) Choice B is correct

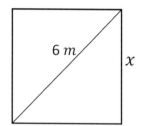

The diagonal of the square is 6 meters. Let x be the side.
Use Pythagorean Theorem: $a^2 + b^2 = c^2$
$x^2 + x^2 = 6^2 \Rightarrow 2x^2 = 6^2 \Rightarrow 2x^2 = 36 \Rightarrow x^2 = 18 \Rightarrow x = \sqrt{18}$
The area of the square is: $\sqrt{18} \times \sqrt{18} = 18 \ m^2$

15) Choice D is correct

If the length of the box is 24, then the width of the box is one third of it, 8, and the height of the box is 4 (half of the width). The volume of the box is: $V = lwh = (24)(8)(4) = 768$

16) Choice D is correct

Write the equation and solve for B: $0.60A = 0.20B$, divide both sides by 0.20, then:

$\frac{0.60}{0.20}A = B$, therefore: $B = 3A$, and B is 3 times of A or it's 300% of A.

17) Choice C is correct

Use simple interest formula: $I = prt$ ($I = interest$, $p = principal$, $r = rate$, $t = time$)

$I = (8,000)(0.025)(5) = 1,000$

18) Choice B is correct

Use percent formula: $part = \frac{percent}{100} \times whole$. $15 = \frac{percent}{100} \times 20 \Rightarrow 15 = \frac{percent \times 20}{100} \Rightarrow$

$15 = \frac{percent \times 2}{10}$, multiply both sides by 10. $150 = percent \times 2$, divide both sides by 2.
$75 = percent$

19) Choice B is correct

Add the first 5 numbers. $40 + 45 + 50 + 35 + 55 = 225$

To find the distance traveled in the next 5 hours, multiply the average by number of hours.

Distance = Average × Rate = $45 \times 5 = 225$,   Add both numbers. $225 + 225 = 450$

20) Choice B is correct

Use distance formula: $Distance = Rate \times time \Rightarrow 420 = 60 \times T$, divide both sides by 60.

$\frac{420}{60} = T \Rightarrow T = 7 \ hours$

21) Choice D is correct.

Plug in each pair of numbers in the equation. The answer should be 20.

A. $(2, 1)$: $4(2) + 6(1) = 14$ No!

B. $(-1, 3)$: $4(-1) + 6(2) = 8$ No!
C. $(-2, 2)$: $4(-2) + 6(2) = 4$ No!
D. $(2, 2)$: $4(2) + 6(2) = 20$ Yes!

22) Choice C is correct

Let x be the number. Write the equation and solve for x.

$\frac{2}{3} \times 15 = \frac{2}{5} \cdot x \Rightarrow \frac{2 \times 15}{3} = \frac{2x}{5}$, use cross multiplication to solve for x. $5 \times 30 = 2x \times 3 \Rightarrow$

$150 = 6x \Rightarrow x = 25$

23) Choice C is correct

To find the discount, multiply the number by $(100\% - rate\ of\ discount)$.

Therefore, for the first discount we get: $(D)(100\% - 20\%) = (D)(0.80) = 0.80\ D$

For increase of 15%: $(0.80D)(100\% + 15\%) = (0.80\ D)(1.15) = 0.92\ D = 92\%\ of\ D$

24) Choice B is correct

Use the formula for Percent of Change: $\frac{\text{New Value} - \text{Old Value}}{Old\ Value} \times 100\%$

$\frac{28 - 45}{45} \times 100\% = -37.7\%$ (negative sign here means that the new price is less than old price).

25) Choice B is correct

1 meter = 100 centimeters. Then: 27,356 × 0.01 = 273.56

26) Choice B is correct

If the score of Mia was 60, therefore the score of Ava is 30. Since, the score of Emma was half as that of Ava, therefore, the score of Emma is 15.

27) Choice D is correct

If 20 balls are removed from the bag at random, there will be one ball in the bag. The probability of choosing a white ball is 1 out of 21. Therefore, the probability of not choosing a white ball is 20 out of 21 and the probability of having not a white ball after removing 20 balls is the same.

28) Choice D is correct

$8 × 10 = $80,   Petrol use: $10 × 2 = 20$ liters,   Petrol cost: $20 × $1 = 20

Money earned: $80 − $20 = 60

29) Choice C is correct

Let x be the original price. If the price of the sofa is decreased by 15% to $476, then: $85\%\ of\ x =$
$476 \Rightarrow 0.85x = 476 \Rightarrow x = 476 \div 0.85 = 560$

30) Choice C is correct

Let x be the number. Write the equation and solve for x. $(28 - x) \div x = 3$

Multiply both sides by x. $(28 - x) = 3x$, then add x both sides. $28 = 4x$, now divide both sides by 4. $x = 7$

PERT Mathematics Practice Test 2

Answers and Explanations

1) Choice D is correct

Let a and b be the numbers. Then: $a + b = x$. $a = 9 \rightarrow 9 + b = x \rightarrow b = x - 9$

$2b = 2(x - 9)$

2) Choice A is correct

Write a proportion and solve for the missing number. $\frac{32}{12} = \frac{6}{x} \rightarrow 32x = 6 \times 12 = 72$

$32x = 72 \rightarrow x = \dfrac{72}{32} = 2.25$

3) Choice A is correct

To solve absolute values equations, write two equations. $2x - 6$ can equal positive 12, or negative 12. Therefore, $2x - 6 = 12 \Rightarrow 2x = 18 \Rightarrow x = 9$.

$2x - 6 = -12 \Rightarrow 2x = -12 + 6 = -6 \Rightarrow x = -3$.

Find the product of solutions: $-3 \times 9 = -27$

4) Choice B is correct

The equation of a line in slope intercept form is: $y = \mathrm{m}x + b$. Solve for y. $3x - y = 6 \rightarrow$

$-y = -3x + 6$. Divide both sides by (-1). Then: $-y = -3x + 6 \rightarrow y = 3x - 6$

The slope of this line is 3. The product of the slopes of two perpendicular lines is -1. Therefore, the slope of a line that is perpendicular to this line is:

$$m_1 \times m_2 = -1 \Rightarrow 3 \times m_2 = -1 \Rightarrow m_2 = \frac{-1}{3} = -\frac{1}{3}$$

5) Choice C is correct

Plug in the value of x and y. $3(x - 2y) + (2 - x)^2$ when $x = 5$ and $y = -3$

$3(x - 2y) + (2 - x)^2 = 3(5 - 2(-3)) + (2 - 5)^2 = 3(5 + 6) + (-3)^2 = 33 + 9 = 42$

6) Choice C is correct

$$1\ yard = 12\ feet$$

$$\frac{(15\ feet\ +\ 7\ yards)}{4} = \frac{(15\ feet\ +\ 21\ feet)}{4} = \frac{(36\ feet\ )}{4} = 9\ feet$$

7) Choice D is correct

Solve for x. $x - 4 \leq 4x - 8 < 16 \Rightarrow$ (add 8 all sides) $-4 + 8 < 4x - 8 + 8 < 16 + 8 \Rightarrow$

$4 < 4x < 24 \Rightarrow$ (divide all sides by 4) $1 \leq x < 6$

x is between 1 and 6. Choice D represents this inequality.

8) Choice D is correct

$Volume\ of\ a\ box\ =\ length \times width \times height = \ 4 \times 5 \times 6 = 120$

9) Choice A is correct

Simplify and combine like terms. $(6x^3 - 8x^2 + 2x^4) - (4x^2 - 2x^4 + 2x^3) \Rightarrow$
$(6x^3 - 8x^2 + 2x^4) - 4x^2 + 2x^4 - 2x^3 \Rightarrow 4x^4 + 4x^3 - 12x^2$

10) Choice C is correct

the population is increased by 15% and 20%. 15% increase changes the population to 115% of original population. For the second increase, multiply the result by 120%.

$(1.15) \times (1.20) = 1.38 = 138\%$. 38 percent of the population is increased after two years.

11) Choice C is correct

Three times of 24,000 is 72,000. One sixth of them cancelled their tickets.

One sixth of 72,000 equals 12,000 ($\frac{1}{6} \times 72,000 = 12,000$).

60,000 ($72000 - 12,000 = 60,000$) fans are attending this week

12) Choice D is correct

$\dfrac{7}{25} = 0.28$

13) Choice D is correct

$\dfrac{18}{24} = \dfrac{3}{4}$

14) Choice C is correct

$average\ (mean)\ = \dfrac{sum\ of\ terms}{number\ of\ terms} \Rightarrow 88 = \dfrac{sum\ of\ terms}{50} \Rightarrow sum = 88 \times 50 = 4,400$

The difference of 94 and 69 is 25. Therefore, 25 should be subtracted from the sum.

$4400 - 25 = 4,375$, mean $\dfrac{\text{sum of terms}}{\text{number of terms}} \Rightarrow mean = \dfrac{4,375}{50} = 87.5$

15) Choice C is correct

All angles in a triangle sum up to 180 degrees. $53 + 45 = 98$. $180 - 98 = 82$,

The third angle is 82 degrees.

16) Choice B is correct

The diagonal of the square is 8. Let x be the side.

Use Pythagorean Theorem: $a^2 + b^2 = c^2$

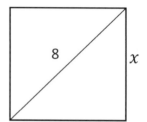

$x^2 + x^2 = 8^2 \Rightarrow 2x^2 = 8^2 \Rightarrow 2x^2 = 64 \Rightarrow x^2 = 32 \Rightarrow x = \sqrt{32}$

The area of the square is: $\sqrt{32} \times \sqrt{32} = 32$

17) Choice B is correct

$\text{Probability} = \dfrac{number\ of\ desired\ outcomes}{number\ of\ total\ outcomes} = \dfrac{18}{12+18+18+24} = \dfrac{18}{72} = \dfrac{1}{4}$

18) Choice A is correct

The width of the rectangle is twice its length. Let x be the length. Then, $width = 2x$

Perimeter of the rectangle is $2\ (width + length) = 2(2x + x) = 60 \Rightarrow 6x = 60 \Rightarrow x = 10$

Length of the rectangle is 10 meters.

19) Choice D is correct

$\text{average} = \dfrac{\text{sum of terms}}{\text{number of terms}} \Rightarrow$ (average of 6 numbers) $12 = \dfrac{\text{sum of numbers}}{6} \Rightarrow$ sum of 6 numbers is $12 \times 6 = 72$

(average of 4 numbers) $10 = \dfrac{\text{sum of numbers}}{4} \Rightarrow$ sum of 4 numbers is $10 \times 4 = 40$

$sum\ of\ 6\ numbers - sum\ of\ 4\ numbers = sum\ of\ 2\ numbers, 72 - 40 = 32$,

average of 2 numbers $= \dfrac{32}{2} = 16$

20) Choice C is correct

Solving Systems of Equations by Elimination

Multiply the first equation by (-2), then add it to the second equation.

$\begin{matrix} -2(2x + 5y = 11) \\ 4x - 2y = -14 \end{matrix} \Rightarrow \begin{matrix} -4x - 10y = -22 \\ 4x - 2y = -14 \end{matrix} \Rightarrow -12y = -36 \Rightarrow y = 3$

Plug in the value of y into one of the equations and solve for x.

$2x + 5(3) = 11 \Rightarrow 2x + 15 = 11 \Rightarrow 2x = -4 \Rightarrow x = -2$

21) Choice B is correct

The perimeter of the trapezoid is $36 \ cm$.

Therefore, the missing side (height) is = $36 - 8 - 12 - 6 = 10$

Area of a trapezoid: $A = \frac{1}{2} h (b_1 + b_2) = \frac{1}{2} (10) (6 + 8) = 70$

22) Choice B is correct

The probability of choosing a Hearts is $\frac{13}{52} = \frac{1}{4}$

23) Choice C is correct

Th ratio of boy to girls is $4:7$. Therefore, there are 4 boys out of 11 students. To find the answer, first divide the total number of students by 11, then multiply the result by 4. $44 \div 11 = 4 \Rightarrow 4 \times 4 = 16$. There are 16 boys and $28 (44 - 16)$ girls. So, 12 more boys should be enrolled to make the ratio $1:1$

24) Choice A is correct

2,500 out of 55,000 equals to $\frac{2500}{55000} = \frac{25}{550} = \frac{1}{22}$

25) Choice D is correct

Isolate and solve for x. $\frac{2}{3}x + \frac{1}{6} = \frac{1}{3} \Rightarrow \frac{2}{3}x = \frac{1}{3} - \frac{1}{6} = \frac{1}{6} \Rightarrow \frac{2}{3}x = \frac{1}{6}$

Multiply both sides by the reciprocal of the coefficient of x. $(\frac{3}{2})\frac{2}{3}x = \frac{1}{6}(\frac{3}{2}) \Rightarrow x = \frac{3}{12} = \frac{1}{4}$

26) Choice B is correct

Use simple interest formula: $I = prt$ (I = interest, p = principal, r = rate, t = time)

$I = (12000)(0.035)(2) = 840$

27) Choice D is correct

Simplify. $6x^2y^3(2x^2y)^3 = 6x^2y^3(8x^6y^3) = 48x^8y^6$

28) Choice C is correct

Surface Area of a cylinder $= 2\pi r (r + h)$, The radius of the cylinder is $3 (6 \div 2)$ inches and its height is 8 inches. Therefore, Surface Area of a cylinder $= 2\pi (3) (3 + 8) = 66 \pi$

29) Choice B is correct

Write the numbers in order: $2, 19, 27, 28, 35, 44, 67$.

Median is the number in the middle. So, the median is 28.

30) Choice D is correct

First, notice that the line $-6y = 16x - 12$ has a negative slope. (graph the line and see that the line has a negative slope) Then, Choices B and C that show the line with positive slope are incorrect.

Now, choose the testing point $(0,0)$ and plug in the values of x and y in the inequality.

$(0,0) \rightarrow -6y \leq 16x - 12 \rightarrow -6(0) \leq 16(0) - 12 \rightarrow 0 \leq -12$

Number 0 is not less than -12. Then, the testing point $(0,0)$ is not in the solution section.

Therefore, choices A and B are incorrect. (notice that both graphs of A and B show the testing point $(0,0)$ in the solution section.)

Only choice D represents the inequality $-6y \leq 16x - 12$

www.EffortlessMath.com

... So Much More Online!

✓ FREE Math lessons

✓ More Math learning books!

✓ Mathematics Worksheets

✓ Online Math Tutors

Need a PDF version of this book?

Visit www.EffortlessMath.com

Receive the PDF version of this book or get another FREE book!

Thank you for using our Book!

Do you LOVE this book?

Then, you can get the PDF version of this book or another book absolutely FREE!

Please email us at:

info@EffortlessMath.com

for details.

Author's Final Note

I hope you enjoyed reading this book. You've made it through the book! Great job!

First of all, thank you for purchasing this study guide. I know you could have picked any number of books to help you prepare for your PERT Math test, but you picked this book and for that I am extremely grateful.

It took me years to write this study guide for the PERT Math because I wanted to prepare a comprehensive PERT Math study guide to help test takers make the most effective use of their valuable time while preparing for the test.

After teaching and tutoring PERT math for over a decade, I've gathered my personal notes and lessons to develop this study guide. It is my greatest hope that the lessons in this book could help you prepare for your test successfully.

If you have any questions, please contact me at reza@effortlessmath.com and I will be glad to assist. Your feedback will help me to greatly improve the quality of my books in the future and make this book even better. Furthermore, I expect that I have made a few minor errors somewhere in this study guide. If you think this to be the case, please let me know so I can fix the issue as soon as possible.

If you enjoyed this book and found some benefit in reading this, I'd like to hear from you and hope that you could take a quick minute to post a review on the book's Amazon page. To leave your valuable feedback, please visit: amzn.to/3h2uOND

Or scan this QR code.

I personally go over every single review, to make sure my books really are reaching out and helping students and test takers. Please help me help PERT Math test takers, by leaving a review!

I wish you all the best in your future success!

Reza Nazari

Math teacher and author

Made in the USA
Columbia, SC
27 August 2021